CURRICULUM AND EVALUATION
S T A N D A R D S
FOR SCHOOL MATHEMATICS
ADDENDA SERIES, GRADES 9–12

WITHDRAWN

ALGEBRA IN A TECHNOLOGICAL WORLD

M. Kathleen Heid

with

Jonathan Choate

Charlene Sheets

Rose Mary Zbiek

Consultants

Harold L. Schoen

Daniel Teague

Christian R. Hirsch, Series Editor

NATIONAL COUNCIL OF
TEACHERS OF MATHEMATICS

Copyright © 1995 by
THE NATIONAL COUNCIL OF TEACHERS OF MATHEMATICS, INC.
1906 Association Drive, Reston, VA 20191-1593
All rights reserved

Third printing 1997

Library of Congress Cataloging-in-Publication Data

Heid, Mary Kathleen.
    Algebra in a technological world / M. Kathleen Heid, with Jonathan
Choate, Charlene Sheets, Rose Mary Zbiek ; consultants, Harold L.
Schoen, Daniel Teague.
        p.   cm. — (Curriculum and evaluation standards for school
mathematics addenda series. Grades 9–12)
    Includes bibliographical references.
    ISBN 0-87353-326-7 (pbk.)
    1. Algebra—Study and teaching (Secondary) I. Title.
II. Series.
QA 159.H45 1995
512.9—dc20   95-10254
                                                    CIP

Printed in the United States of America

ii

TABLE OF CONTENTS

FOREWORD

As the *Curriculum and Evaluation Standards for School Mathematics* (NCTM 1989) was being developed, it became apparent that supporting publications would be needed to aid in interpreting and implementing the curriculum and evaluation standards and the underlying instructional themes. A Task Force on the Addenda to the *Curriculum and Evaluation Standards for School Mathematics,* chaired by Thomas Rowan and composed of Joan Duea, Christian Hirsch, Marie Jernigan, and Richard Lodholz, was appointed by Shirley Frye, then NCTM president, in the spring of 1988. The Task Force's recommendations on the scope and nature of the supporting publications were submitted in the fall of 1988 to the Educational Materials Committee, which subsequently framed the Addenda Project for NCTM Board approval.

Central to the Addenda Project was the formation of three writing teams to prepare a series of publications targeted at mathematics education in grades K–6, 5–8, and 9–12. The writing teams consisted of classroom teachers, mathematics supervisors, and university mathematics educators. The purpose of the series was to clarify the recommendations of selected standards and to illustrate how the standards could realistically be implemented in K–12 classrooms in North America.

The themes of problem solving, reasoning, communication, and connections have been woven throughout each volume in the series. The use of technological tools and the view of assessment as a means of guiding instruction are integral to the publications. The materials have been field-tested by teachers to ensure that they reflect the realities of today's classrooms and to make them "teacher friendly."

We envision the Addenda Series being used as a resource by individuals as they begin to implement the recommendations of the *Curriculum and Evaluation Standards.* In addition, volumes in a particular series would be appropriate for in-service programs and for preservice courses in teacher education programs.

On behalf of the National Council of Teachers of Mathematics, I would like to thank the authors, consultants, and editor, who gave willingly of their time, effort, and expertise in developing these exemplary materials. Special thanks are due to Frank Demana and Bert Waits who contributed to the planning of this book and provided ideas for problem situations. Gratitude is also expressed to Beth Ritsema and Rebecca Walker, who reviewed and commented extensively on drafts of the material as this volume progressed. Finally, the continuing technical assistance of Cynthia Rosso and the able production staff in Reston is gratefully acknowledged.

Bonnie H. Litwiller
Addenda Project Coordinator

PREFACE

The *Curriculum and Evaluation Standards for School Mathematics*, released in March 1989 by the National Council of Teachers of Mathematics, has focused national attention on a new set of goals and expectations for school mathematics. That visionary document provides a broad framework for what the mathematics curriculum in grades K–12 should include in terms of content priority and emphasis. It suggests not only what students should learn but also how that learning should occur and be assessed and evaluated.

Although the *Curriculum and Evaluation Standards* specifies the key components of a quality contemporary school mathematics program, it encourages local initiatives in realizing the vision set forth. In so doing, it offers school districts, mathematics departments, and classroom teachers new opportunities and challenges. The purpose of this volume, and others in the Addenda series, is to provide instructional ideas and materials that will support implementation of the *Curriculum and Evaluation Standards* in local settings. It addresses in a very practical way the content, pedagogy, and pupil assessment dimensions of reshaping school mathematics. As the final volume in the high school series, *Algebra in a Technological World* also projects the future of algebra in classrooms where today's learning tools, such as graphing calculators with table-building, graphing, and function-fitting capabilities, are augmented with the capability to perform symbolic manipulations.

Reshaping Content

The curriculum standards for grades 9–12 identify a common core of mathematical topics that *all* students should have the opportunity to learn. The need to prepare students for the workplace, for college, and for citizenship is reflected in a broad mathematical sciences curriculum. The traditional strands of algebra, functions, geometry, and trigonometry are balanced with topics from data analysis and statistics, probability, and discrete mathematics. The broadening of the content of the curriculum is accompanied by a broadening of its focus. Narrow curricular expectations of memorizing isolated facts and procedures and becoming proficient with by-hand calculations and manipulations give way to developing mathematics as a connected whole with an emphasis on conceptual understanding, multiple representations and their linkages, mathematical modeling, and problem solving.

This broadening of curricular intent is clearly reflected in the grades 9–12 standards for algebra and functions. Together these standards signal a shift in perspective from algebra as skills for transforming, simplifying, and solving symbolic expressions to algebra as a way to express and analyze relationships. Accompanying these standards are suggestions for more integration across topics at all grade levels and for increased attention to real-world applications, matrices, and the use of emerging calculator and computer technologies as tools for problem solving and conceptual development. Among the topics to receive decreased attention are by-hand methods for simplifying radical and rational expressions, for factoring polynomials, and for evaluating and graphing functions.

Algebra in a Technological World brings to life and clothes these recommendations by describing, and illustrating with specific examples, how school algebra can be organized naturally around the concepts of functions, families of functions, and mathematical modeling. Chapter 1 gives an overview of how technological tools such as graphing calculators and computer-algebra systems support new conceptions of algebra that focus on enabling students to explore, describe, and explain quantitative relations in their world. Chapters 3 and 4 elaborate in detail a modeling and functions approach to algebra. Connections between the algebra strand and the geometry and discrete mathematics strands by means of matrices are examined in chapter 5. The final chapter illuminates distinctions among, and suggests priorities that be given to the development of, symbol sense, symbolic manipulation, and symbolic reasoning in a technological world.

A special *Try This* feature appearing throughout the volume furnishes exercises, problems, and explorations for use with students. We hope that these will pique your interest and that you will use the margins productively. More extensive investigations and projects appear as blackline masters at the end of chapters. These activities are appropriate for students at varied levels. Solutions for these activities appear in the appendix. An annotated list of computer resources supporting a modeling and functions approach to algebra is provided at the end of this volume.

Reshaping Pedagogy

The *Curriculum and Evaluation Standards* paints mathematics as an activity and a process, not simply as a body of content to be mastered. Throughout, there is an emphasis on doing mathematics, recognizing connections, and valuing the enterprise. Hence, standards are presented for Mathematics as Problem Solving, Mathematics as Communication, Mathematics as Reasoning, and Mathematical Connections. The intent of these four standards is to frame a curriculum that ensures the development of broad mathematical power in addition to technical competence; that cultivates students' abilities to explore, conjecture, reason logically, formulate and solve problems, and communicate mathematically; and that fosters the development of self-confidence.

Realization of these process and affective goals will require, in many cases, new teaching-learning environments. The traditional view of the teacher as authority figure and dispenser of information must give way to that of the teacher as catalyst and facilitator of learning. To this end, the standards for grades 9–12 call for increased attention to—

♦ actively involving students in constructing and applying mathematical ideas;

♦ using problem solving as a means as well as a goal of instruction;

♦ promoting student interaction through the use of effective questioning techniques;

♦ using a variety of instructional formats—small cooperative groups, individual explorations, whole-class instruction, and projects;

♦ using calculators and computers as tools for learning and doing mathematics.

At the heart of changing patterns of instruction are the increasing capabilities of technology. The standards for grades 9–12 assume that students will have access to graphing calculators and that computers will be available for demonstration purposes as well as for individual and group work. Chapter 2 of *Algebra in a Technological World* examines carefully how the new methodologies and technologies interact to support new curricular goals.

The classroom-ready activity sheets at the end of chapters of this volume furnish tasks that require students to experiment, collect data, search for patterns, make conjectures, and construct mathematical models and improve their features. These activities are ideally suited to cooperative-group work. As students complete these activities, they have multiple opportunities to communicate about important mathematical ideas. As with all student investigations, it is important that provisions be made for students to share their experiences, clarify their thinking, generalize their discoveries where appropriate, and recognize connections with other topics. The previously mentioned *Try This* feature appearing throughout the volume furnishes more structured tasks that also offer opportunities for genuine problem solving, reasoning, and lively classroom discourse.

Teaching Matters is another special feature of this book. These captioned margin notes supply helpful instructional suggestions, including ideas on motivation and on the use of appropriate technological tools. They also identify possible difficulties that students might encounter when exploring certain topics and suggest how these can be anticipated and addressed in instruction.

Reshaping Assessment

Complete pictures of classrooms in which the *Curriculum and Evaluation Standards* is being implemented not only show changes in mathematical content and instructional practice but also reflect changes in the purpose and methods of student assessment. Classrooms where students are expected to be engaged in mathematical thinking and in constructing and reorganizing their own knowledge require adaptive teaching informed by observing and listening to students at work. Thus, informal, performance-based assessment methods are essential to the new vision of school mathematics.

Analysis of students' written work remains important. However, single-answer paper-and-pencil tests are often inadequate to assess the development of students' abilities to analyze and solve problems, make connections, reason mathematically, and communicate mathematically. Potentially richer sources of information include student-produced analyses of problem situations, solutions to problems, reports of investigations, and journal entries. Moreover, if calculator and computer technologies are now to be accepted as part of the environment in which students learn and do mathematics, these tools should also be available to students in most assessment situations.

Algebra in a Technological World reflects the multidimensional aspects of student assessment and the fact that assessment is integral to instruction. Chapter 2 illustrates how function-family exploration tasks can be used for the purpose of guiding instruction and, with an appropriate scoring rubric, for student evaluation. Ways in which interviews can be used to probe further the depth and breadth of student understanding are also exemplified. *Assessment Matters,* a special margin feature, supplies additional suggestions for assessment techniques and tasks related to the content under discussion.

Conclusion

The standards on algebra and functions for grades 9–12 were influenced by the capabilities of emerging graphing calculator and computer technology. This technology supports a vision of school algebra that focuses on conceptual understanding, symbol sense, and mathematical modeling. As amply illustrated in this volume, refocusing algebra on modeling places functions and variables at center stage, where they ought to be. It also serves to break down artificial barriers between algebra and statistics, between algebra and geometry, and between algebra and discrete mathematics. Finally, modeling authentic problems permits and encourages a diversity of approaches thereby promoting greater access by a broader population to meaningful mathematics.

Sustainable change must occur first in the hearts, minds, and classrooms of individual teachers, and then in their departments and school districts. For the kinds of changes envisioned in the *Standards* to be realized, it is essential that parents and school boards understand and support the changing nature of mathematics, particularly algebra in a technological world. Individually we can initiate the process of change; collectively we can make the vision of the *Curriculum and Evaluation Standards* a reality. We hope you will find the information in this book valuable in translating the vision of the *Standards* into practice in your community.

Christian R. Hirsch, Editor
Grades 9–12 Addenda Series

**CHAPTER 1
THE FUTURE OF ALGEBRA
IN A TECHNOLOGICAL WORLD**

NEW VIEWS OF ALGEBRA

Perhaps more than any other area of school mathematics, the study of algebra is bound to change dramatically with the infusion of currently available and emerging technology. What was once the inviolable domain of paper-and-pencil manipulative algebra is now within easy reach of school-level computing technology. This technology demands new visions of school algebra that shift the emphasis away from symbolic manipulation toward conceptual understanding, symbol sense, and mathematical modeling.

No longer can the main purpose of algebra be the fine-tuning of techniques for by-hand symbolic manipulation or the acquisition of a predefined set of procedures for solving a fixed set of problems. Lesson after lesson of "simplify these expressions" or "solve these equations" will no longer characterize the school algebra experience. Students will spend far less time on many of these techniques, will execute a majority of them with computing technology, and will completely forgo the study of others. Although some of the attention now paid to symbolic representations will be rededicated to developing "symbol sense," most class time will be spent in helping students develop a sense for how algebra can be used to explain the world around them. Applications of algebra will no longer be synonymous with "age," "coin," "mixture," and "distance-rate-time" word problems. Students will leave their school algebra experience with answers to such questions as "What good is algebra?"

The language of technology quite naturally depends on the concepts of *variable* and *function*. But the concepts of variable and function in a technological world are much richer than those found in current school textbooks or in the minds of today's students. The search for variable values that satisfy equations need no longer be the unquestioned and primary goals of beginning algebra. Functions are no longer merely abstract objects that "pass the vertical line test." In a technological world, variables actually vary and functions describe real-world phenomena. Variables represent quantities that change, and algebra is the study of relationships among these changing quantities. What was the search for fixed values that fit statically defined relationships is now the dynamic exploration of mathematical relationships.

Not only does technology suggest an increased "front stage" role for functions, but it also allows for the dynamic study of families of functions. Even at the earliest stages of their encounters with algebra, students can examine the effects of changing parameters on the numerical and graphical representations of functions. For example, they can conduct investigations of how changing the values of the parameters a, b, and c affect the shape of the graph of $f(x) = ax^2 + bx + c$. In fact, algebra in a technological world is, more so than ever before, a language of representation. Technology allows students to study algebra as the meaningful and related representations of functions, variables, and relations rather than as the acquisition of skills in manipulating symbolic representations stripped of other meaning.

THE IMPACT OF TECHNOLOGY ON THE CONTENT OF ALGEBRA

The content of school algebra must be different in a technological world. Students now have free access to a variety of graphing tools, to symbolic-manipulation programs, and to spreadsheets of ever-increasing sophistication. They also have access to computing tools, such as computer-algebra systems and multiply linked representational tools, that allow operating on, and communicating among, representations of numerous varieties. These tools are sometimes placed in the context of a game that requires applying particular mathematical concepts. In addition to powerful computing tools, the technological world offers new microworlds for students to explore through videodisc technology or computer-generated simulations of realistic situations. The following sections will discuss examples of how each of these tools can fundamentally change the content of school algebra.

Graphing Tools

Some computing tools and programs will be more influential than others in contributing to a new vision of algebra. One of the most important tools so far in reformulating school algebra has been the function, or relation, grapher. Graphing tools are available in everything from the most powerful supercomputer or mainframe to widely available microcomputers and calculators. Graphing tools influence the content of algebra in a technological world in the following ways:

◆ They allow a ready visualization of relationships.

◆ They allow the accurate solution of equations and inequalities not possible through symbolic manipulation alone.

◆ They provide numerical and graphical solutions that support solutions found using algebraic manipulation.

◆ They promote exploration by students and their understanding of the effect of change in one representation on another representation.

◆ They encourage the exploration of relationships and mathematical concepts.

◆ They promote "what if" modeling of realistic situations.

The following pages outline specific classroom problems that highlight these features of graphing tools and their impact on algebra in a technological world.

Assessment Matters: The many ways that graphing tools influence the content should be reflected in any assessment plan. Assessment activities can be designed that are similar to the instructional activities in this book. A teacher can then assess students' performance on these activities in terms of their ability to visualize relationships, to solve equations and inequalities accurately using graphical methods, and so on, down this list of content influences.

Graphing tools allow a ready visualization of relationships.

Graphs of revenue, cost, and profit functions afford interesting insights into the effects of changing cost conditions. The graphs give an overall picture of relationships, and the scan-and-zoom capacities of calculator or computer graphers help find answers to a given level of precision.

Season-Ticket Sales

A new professional team is in the process of determining the optimal price for a special ticket package for its first season. A survey of potential fans reveals how much they are willing to pay for a four-game package. The data from the survey are displayed in the chart.

Price of the Four-Game Package	Number of Packages That Could Be Sold at That Price
$96.25	5 000
90.00	10 000
81.25	15 000
56.25	25 000
50.00	27 016
40.00	30 000
21.25	35 000

1. On the basis of the foregoing data, find a relationship that describes the price of a package as a function of the number sold (in thousands).

2. Determine the number of ticket packages that must be sold to maximize revenue. What package price will maximize revenue?

3. Suppose that fixed costs for the sales of ticket packages are $200 000 no matter how many are sold. Suppose further that for every thousand packages sold, the costs increase by $23 000. How many packages must be sold to maximize profit? What package price will maximize profit?

4. If fixed costs are lowered to $150 000, determine the effect on the maximum profit.

5. Assuming that fixed costs remain at $200 000, how will an increase in variable cost to $30 for each package affect the maximum profit? The optimal price?

The revenue function can be found through the use of a computer or calculator curve fitter. In this example, a curve fitter was used to find the best-fitting quadratic function for the price of a package as a function of the number of ticket packages sold (in thousands). The maximum revenue can then be found by using a scan-and-zoom procedure on a graph like that in figure 1.1. The maximum profit can be investigated through examining the differences in the graphs of the cost and revenue functions. The revenue function, R, in figure 1.1 is derived from the fitted quadratic by rounding to whole-number coefficients.

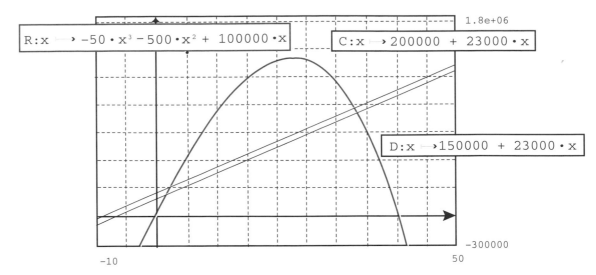

$$R: x \longrightarrow -50 \cdot x^3 - 500 \cdot x^2 + 100000 \cdot x$$

$$C: x \longrightarrow 200000 + 23000 \cdot x$$

$$D: x \longrightarrow 150000 + 23000 \cdot x$$

1.8e+06

−300000

−10

50

Fig. 1.1. Graphs of revenue, R, the first cost, C, and the second cost, D, as functions of x, the number of thousands of ticket packages sold in the season-ticket situation

Whereas an increase in fixed costs results in a cost function that is parallel to the original, an increase in the variable cost results in a cost function, *G,* with a greater slope, like that shown in figure 1.2.

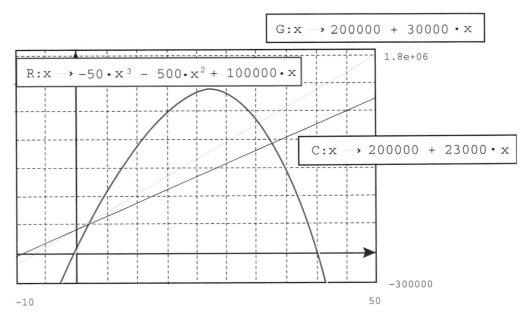

$$G: x \longrightarrow 200000 + 30000 \cdot x$$

$$R: x \longrightarrow -50 \cdot x^3 - 500 \cdot x^2 + 100000 \cdot x$$

$$C: x \longrightarrow 200000 + 23000 \cdot x$$

1.8e+06

−300000

−10

50

Fig. 1.2. Graphs of revenue, R, the first cost, C, and the third cost, G, as functions of x, the number of thousands of packages sold in the season-ticket situation

A variety of geometrical problems lend themselves to algebraic-graphical representation when graphical computing tools are available. Consider the following graphical exploration of a classic maximization problem.

Figures 1.3a, 1.3b, and 1.3c show how a cone is constructed and how a formula for its radius is derived. Figure 1.3d illustrates a right triangle embedded in the cone.

The Volume of a Cone

A cone is constructed from a flat circular piece of material six centimeters in diameter by removing a sector of arc length x centimeters and then connecting the edges as shown in the sequence of diagrams in figure 1.3. Find the arc length that will result in a cone with the largest possible volume.

Teaching Matters: Have students construct cones by this process. Do the physical models support these statements? Have students compare models with the graphical display to support their graphical interpretation.

(a)

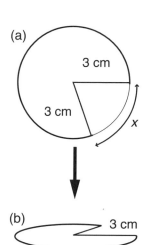

The circumference of the remaining sector equals $6\pi - x$, where x is the arc length of the sector removed.

(b)

3 cm

(c)

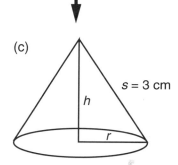

The radius of the circular base of the resulting cone is r. The circumference of that base is $6\pi - x$, the same as the circumference of the remaining sector.

So $2\pi r = 6\pi - x$; solving for r gives

$$r = \frac{6\pi - x}{2\pi}.$$

(d)

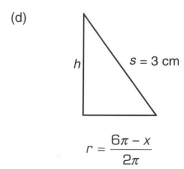

$s = 3$ cm

$$r = \frac{6\pi - x}{2\pi}$$

Fig. 1.3. Derivation of a formula for the radius of the cone

When the Pythagorean theorem is applied,

$$s^2 = h^2 + r^2$$

or

$$3^2 = h^2 + \left(\frac{6\pi - x}{2\pi}\right)^2,$$

so

$$h = \sqrt{9 - \left(\frac{6\pi - x}{2\pi}\right)^2}.$$

Since the volume of the cone is one-third the area of the base times the height,

$$V(x) = \left(\frac{1}{3}\right)\pi r^2 h = \frac{\pi}{3}\left(\frac{6\pi - x}{2\pi}\right)^2 \sqrt{9 - \left(\frac{6\pi - x}{2\pi}\right)^2},$$

which is a function of the arc length, x, of the removed sector.

The graph (see fig. 1.4) gives an overall picture of the relationship between the volume and the size of the piece removed. We can see at a glance that the maximum volume occurs for an x-value between approximately 3 and 4. We can also see that the volume increases rapidly for values of x from 0 to the x-value at which the maximum occurs but decreases slowly thereafter.

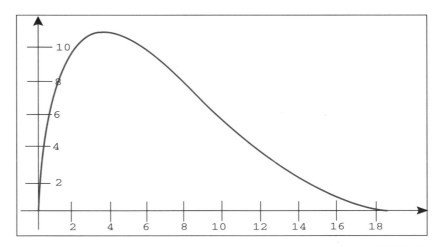

Fig. 1.4. Graph of $V(x) = \left(\frac{1}{3}\right)\pi r^2 h = \frac{\pi}{3}\left(\frac{6\pi - x}{2\pi}\right)^2 \sqrt{9 - \left(\frac{6\pi - x}{2\pi}\right)^2}$

Technology not only enables algebra students to visualize functions of a single variable but also makes accessible the visualization of functions of two variables. At an early stage in their exposure to algebraic ideas, students can make sense of functions of two variables through graphical representations. Using powerful computing tools, students can develop intuitive understandings about some fundamental notions of calculus.

The graph of function G (fig. 1.5) represents the number of gallons of gasoline used each year as a function of the number of miles driven in the city and the number of miles driven on highways. Given that 10 000 miles are driven each year on highways, the darkened line describes the relationship between the number of miles driven in the city and the number of gallons of gasoline used each year. In this particular graph, the darkened line offers an early concrete representation related to the concept of a partial derivative.

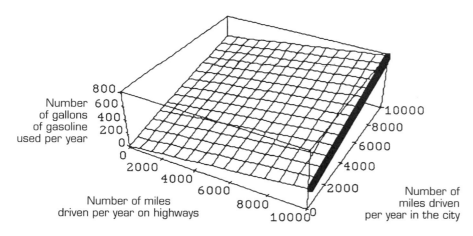

Fig. 1.5. The graph of function G represents the number of gallons of gasoline used each year as a function of the number of miles driven in the city and on highways. The graph was produced using the computer-algebra program Mathematica.

Three-dimensional graphs can also illustrate a variety of the complexities in functional relationships. For example, in figure 1.6 (Sarmiento 1993), several trends can be noted in the three-dimensional graph of the concentration of CO_2 in the atmosphere as a function of time and latitude. In the graph depicting CO_2 concentration from 1986 through 1991, we can see the approximately linear increase in CO_2 concentration across time coupled with its sinusoidal fluctuation with the seasons. Also, the decay in amplitude of these seasonal fluctuations is easy to see as we move from the North Pole to the South Pole. The author of the article from which the graph was taken attributes the sinusoidal fluctuation to variations in the amount of photosynthesis taking place in the terrestrial biosphere; he attributes the decay in amplitude of these seasonal fluctuations to the fact that in the Southern Hemisphere, the terrestrial biosphere covers too small an area to have a significant impact.

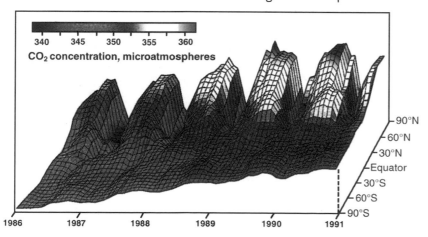

Fig. 1.6. Three-dimensional graph of the concentration of CO_2 in the atmosphere as a function of time and latitude

Graphing tools allow the accurate solution of equations and inequalities not possible through symbolic manipulation alone.

Consider the task of finding to the nearest 0.01 the two largest values of x that satisfy the equation $\sin(x) = e^x$. Figure 1.7 shows the superimposed graphs of $f(x) = \sin(x)$ and $g(x) = e^x$.

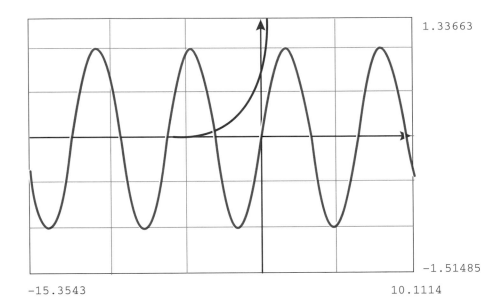

1.33663

−1.51485

−15.3543   10.1114

Fig. 1.7. Computer-generated graphs of f(x) = sin(x) and g(x) = e^x

Although the solution to this equation is inaccessible using standard by-hand symbolic-manipulation methods, by performing a sequence of scan-and-zoom procedures on a function grapher, students can approximate the desired values to acceptable accuracy. They can then concentrate on more global reasoning and observations: an infinite number of negative values solve this equation but no positive values; the meaning of finding a solution to the nearest hundredth needs to be clarified; and results can be verified graphically and numerically.

The interplay between a knowledge of function families and their visual representations comes into view with the analysis of this task. Since the highest two values of x satisfying the equation are approximately −3.18 and −6.28, it is tempting to note the proximity of these two answers to −π and −2π and to surmise some regularity to the solutions. It is clear, however, from the properties of these function families that the increasing nature of e^x and the periodicity of $\sin(x)$ prevent the solutions from being equally spaced. For these same reasons we know that as x decreases, the zeros of the sine function become better and better approximations for the solutions to the equation. As $x \to -\infty$, $e^x \to 0$; the solutions get closer and closer to multiples of π. Since we are looking for the zeros to the nearest 0.01, we *will* find that the two solutions coincide.

Graphing tools provide numerical and graphical solutions that support solutions found using algebraic manipulation.

In the season-ticket situation (see page 3), we might want to determine the break-even price of the ticket package. A symbolic solution could be obtained by solving the cubic equation

$$-50x^3 - 500x^2 + 100\,000x = 200\,000 + 23\,000x.$$

We could examine the graph and determine that for a cost function $C(x) = 200\,000 + 23\,000x$, the higher break-even price is between $30 and $35. A scan-and-zoom procedure will give a more precise value very close to $33, corroborating the decimal approximations shown in figure 1.8 obtained from Maple, a symbolic-manipulation program, the core of Calculus T/L II.

```
P:x → -50•x³ - 500•x² + 77000•x - 200000
```

P(x) = 0 if x is in {x1 , x2 , x3} where

x1 = 32.9970   x2 = 2.65534   x3 = -45.6524

Fig. 1.8. Decimal approximations for the solution to $-50x^3 - 500x^2 + 100\,000x = 200\,000 + 23\,000x$ (or equivalently, $-50x^3 - 500x^2 + 77\,000x - 200\,000 = 0$)

Graphing tools promote exploration by students and their understanding of the effect of change in one representation on another representation.

Consider the following condensed version of a situation and exploration drawn from *Concepts in Algebra: A Technological Approach* (Fey and Heid 1995), which gives a graphical meaning to the principle of the subtraction of equalities. Explorations like these can help students link symbolic procedures with graphical representations.

Copying-Machine Situation

An office manager must decide between two options to fill the copying needs of his department. He wants to find the more economical option.

- Ace Copiers, the first company contacted, offers to lease a copy machine for a fixed weekly fee of $50 and an additional charge of 2.1¢ ($0.021) for each copy.

- For the same machine and comparable service, a second company, Speedy Print, offers a fixed charge of $180 a week with an additional charge of 0.5¢ ($0.005) for each copy.

Students can let the variable n represent the number of copies the office requires during a week and write function rules for the two companies:

Ace Copiers: $f(n) = 0.021n + 50$

Speedy Print: $g(n) = 0.005n + 180$

A graph can be useful in comparing the two pricing structures. We can see from the graphs in figure 1.9 that the two companies charge the same when just over 8000 copies are made. Ace Copiers is less expensive for fewer copies, and Speedy Print is less expensive for more copies.

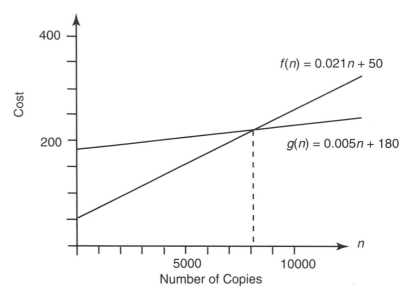

Fig. 1.9. Graphs of price as a function of the number of copies made for Ace Copiers and Speedy Print

Consider the following continuation of the copying-machine situation.

Copying-Machine Situation with a Change in Conditions

To entice customers during the summer season, Ace Copiers decides to eliminate its fixed charge of $50 a week. According to its advertisements, customers pay only for the copies they make! When Speedy Print learns about the impending change, it immediately enters the price war by reducing its fixed charge, also by $50.

The office manager, having analyzed the original fee schedule, wants to know how these reductions affect the relative advantages of the deals available from the two companies.

Because the situation has been slightly modified, the function rules and their representations are also slightly modified. For example, instead of looking at $f(n)$ and $g(n)$, we are looking at

$$h(n) = f(n) - 50 = 0.021n$$

and

$$j(n) = g(n) - 50 = 0.005n + 130.$$

It is interesting to note in figure 1.10 the relationship between the graphical representation of

$$0.021n + 50 = 0.005n + 180$$

and the graphical representation of

$$(0.021n + 50) - 50 = (0.005n + 180) - 50.$$

Although both function graphs shift 50 units down, the intersection of the transformed graphs has the same input value as the original graphs.

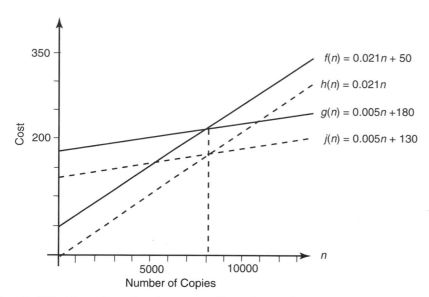

Fig. 1.10. Graphs of price as a function of the number of copies made for Ace Copiers and Speedy Print before and after the conditions changed and the price war began

Graphing tools encourage the exploration of relationships and mathematical concepts.

The existence of computer and calculator graphers allows students to build on their graphical reasoning to enhance their understanding of functions. Dugdale, Wagner, and Kibbey (1992) have described ways to help students reason about polynomial graphs as the sums of monomial graphs. They point out that their goal is not to enhance students' abilities to produce polynomial graphs but rather to help students build "a qualitative understanding of the behavior of polynomial functions and their graphs" (p. 123).

It is interesting to compare the graphs (see fig. 1.11) of $f(x) = -3x^2$ and $g(x) = x^6$ with the graph of $h(x) = x^6 - 3x^2$. We might notice, for example, that the graph of $h(x) = x^6 - 3x^2$ resembles the graph of $f(x) = -3x^2$ for x-values between -1 and 1 and that it resembles the graph of $g(x) = x^6$ elsewhere. This situation makes sense, since $|x^6| < |-3x^2|$ for values of x between -1 and 1.

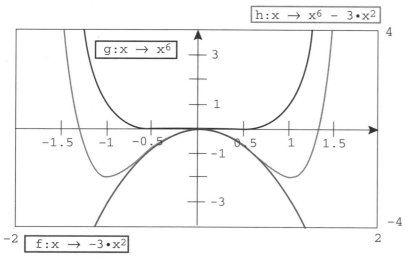

Fig. 1.11. Graphs of $f(x) = -3x^2$, $g(x) = x^6$, and $h(x) = f(x) + g(x) = x^6 - 3x^2$

Activity 1, "The Effects of Monomial Terms on Polynomial Functions," at the end of this chapter illustrates this point of view on the qualitative analysis of graphs. It can be assigned after students have acquired some experience with the shapes of monomial functions and after an example like the one just explained has been explored.

Graphing tools promote "what if" modeling of realistic situations.

Access to a function grapher and curve fitter makes mathematical modeling a dynamic and interactive process. Students can identify variables, conduct experiments, and gather data. They can then use a computer or calculator curve fitter and their knowledge of the situation in their search for an appropriate function rule to fit the data and the situation.

This process of "what if" modeling is exemplified in an experiment included in *Concepts in Algebra: A Technological Approach* (Fey and Heid 1995, pp. 308–11). Students are asked to find a relationship between the height of a ramp and the time it takes a skateboard to travel the length of the ramp (see fig. 1.12).

After students gather the data, they think about the situation, examine the data, and conjecture about the type of function that best describes the situation. Students will need to decide on the type of function rule to try ahead of time, since most computer or calculator curve fitters require the user to select the type of function rule.

In the skateboard experiment, students start their curve fitting by examining data sets like that in figure 1.13 (although many students generate more than four data pairs), along with a plot of the data. The data given here are in a small but reasonable range. In general, many more data points should be used. The table of values, graphs, and fitted curves were created with Calculus T/L II software.

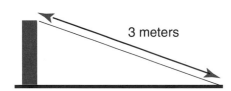

Fig. 1.12. Skateboard-ramp experiment layout

InData	OutData
15	4.4
30	2.9
45	2.5
60	1.6

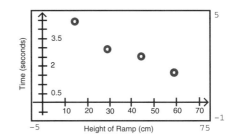

Fig. 1.13. Table and plot of sample data for the skateboard experiment

Such data sets are usually produced by taking several time measurements at each height and using their average to reduce the effects of measurement errors.

As students consider the type of function rule they think will best fit, they take into consideration such questions as these:

◆ What happens to the run time as the ramp gets higher?

◆ What happens to the run time as the ramp gets flatter?

◆ Will the run time ever be 0?

◆ Will the ramp height ever be 0?

Students who consider these questions first may quickly eliminate the linear and quadratic functions. They may reason that the best model probably is not linear because the skateboard will not move when the ramp height is 0. The best model probably is not quadratic because it is unlikely that the run time will both increase and decrease.

Students will differ greatly, however, in the extent to which they focus on reducing the numerical error in the curve fit. Students who start with the four data points in figure 1.13, and who are more attracted to finding a curve that fits the data with the least error, can still be engaged in a discussion of the meaning of particular mathematical models. In this example, since the data appear to be fairly linear, these students may start with a linear rule as shown in figure 1.14.

Although the linear function with the rule

$$f(x) = -0.0586666x + 5.05000$$

fits the data fairly well, students will probably notice that the $f(x)$-intercept of about 5 and the x-intercept of about 85 do not make sense. After all, we would not expect a skateboard to move on its own on a flat ramp, let alone traverse the ramp in about five seconds—as suggested by an $f(x)$-intercept of 5! Moreover, the ramp will take some time to traverse, no matter how high it is, so an x-intercept of any positive value does not make sense in terms of the situation. The intercepts continue to be a problem as students consider best-fitting curves from several different families of functions.

Students who are set on increasing the goodness-of-fit value or decreasing the amount of error in their curve fit are sometimes attracted to increasing the degree of the polynomial. In this example, they find that although a quadratic function fits the data better, it does not fit the situation better because higher ramp heights are not likely to produce longer run times as is the situation (see fig. 1.15) for the best-fitting quadratic function given by

$$f(x) = 0.000666666x^2 - 0.1086666622x + 5.79999.$$

The students' quest continues, with the best-fitting exponential function (see fig. 1.16), given by

$$f(x) = 5.92368(0.9790021063^x),$$

addressing the x-intercept problem but not the $f(x)$-intercept problem.

Here, students may exhaust the choices of function families available on the program menu before they find an appropriate rule. They can then experiment with a few function rules of their own choosing, watching for an improved fit between the function rule and the situation. In figure 1.17, we see a few attempts at fitting a rational function of the form $f(x) = a/x$.

Although a rational function of the form $f(x) = a/x$ models the needed asymptotic behavior along the input and output axes, students may find that a function of the form

$$f(x) = \frac{a}{\sqrt{x}}$$

is an even better model because negative input and output values are eliminated. A detailed discussion of how this experiment plays out with ninth-grade students can be found in Zbiek and Heid (1990). Activity 2,

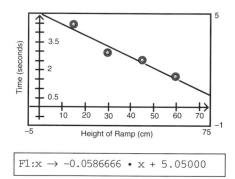

F1: x → −0.0586666 · x + 5.05000

Fig. 1.14. Graph and rule corresponding to the best-fitting linear function for the skateboard experiment sample data

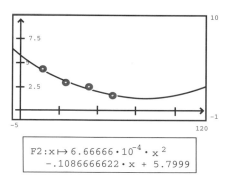

$F2: x \mapsto 6.66666 \cdot 10^{-4} \cdot x^2$
$\quad -.1086666622 \cdot x + 5.7999$

Fig. 1.15. Graph and rule corresponding to the best-fitting quadratic function for the skateboard experiment sample data

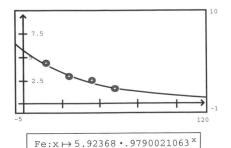

$$Fe:x \mapsto 5.92368 \cdot .9790021063^x$$

Fig. 1.16. Graph and rule corresponding to the best-fitting exponential function for the skateboard experiment sample data

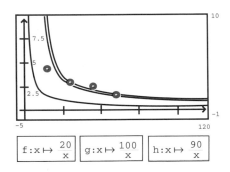

$$f:x \mapsto \frac{20}{x} \qquad g:x \mapsto \frac{100}{x} \qquad h:x \mapsto \frac{90}{x}$$

Fig. 1.17. Graphs and rules corresponding to several rational functions for the skateboard experiment sample data

"Oil and Water Don't Mix," another modeling activity of this sort, is derived from an investigation developed by the Core-Plus Mathematics Project as part of an integrated mathematics curriculum for all students (CPMP 1994).

Symbolic-Manipulation Tools

Symbolic-manipulation tools allow the development of symbol sense.

Whereas graphing tools open up new representational worlds to beginning algebra students, symbolic-manipulation tools allow school mathematics classes to de-emphasize by-hand symbolic manipulation. Instead of concentrating on mastery of routine procedures, students can focus on what the symbols mean. For example, consider the following situation.

> The profit the local high school makes from a concert given by a popular rock band is a function of the price, t, of a ticket where $P(t) = -75t^2 + 1500t - 4800$.

A number of equivalent forms can be quickly produced by a symbolic-manipulation program for the expression

$$-75t^2 + 1500t - 4800.$$

Equivalent expressions include

$$(300 - 75t)(t - 16)$$

and

$$t(-75t + 1500) - 4800.$$

Which expression most readily shows the break-even ticket prices? [The factored form tells that $16 and $4 are the break-even ticket prices.] Which expression most readily shows the fixed cost of holding the concert? [The other two forms readily give the information that the fixed cost is $4800.] What could the expression $t(-75t + 1500)$ mean in terms of the concert situation? [The expression $t(-75t + 1500)$ could represent the revenue from sales of tickets with t the price of a ticket and $-75t + 1500$ the number of tickets that will be sold at the ticket price of t. If $-75t + 1500$ is the number of tickets sold, then for every dollar increase in the price of a ticket, 75 fewer tickets are sold.] Since symbolic-manipulation programs can easily produce these forms, the ability to give meaning to the forms and the ability to identify which forms would be useful in which circumstances become important.

Symbolic-manipulation tools allow concentration on macroprocedures and principles governing symbolic manipulation.

Even when the required focus is the by-hand solution of equations, symbolic-manipulation tools can be used to help students focus on the concept of equivalence and on the properties of equality instead of on the by-hand production of equivalent expressions and equations. In the example in figure 1.18, the student uses a symbolic-manipulation program to give the commands in the left-hand column while the symbolic-manipulation program performs the manipulations.

To issue these commands, students need to recognize the more global features of the current form of the equation and its component expressions.

For example, the command to square both sides of the equation in step 1 suggests a recognition of radical expressions and of an appropriate procedure for solving a radical equation. The command to expand the right-hand side of the equation in step 2 suggests a recognition of the difference between expressions and equations and of the need to produce an equivalent expression. With a symbolic-manipulation program to execute routine symbolic procedures, students can concentrate on more global procedures and principles. It is less important, for example, to be able to execute the squaring procedure than to recognize when such action is appropriate.

Command issued by user	Computer-generated output
	$\sqrt{5 \cdot y + 1} = 1 + \sqrt{3 \cdot y}$
① Square both sides of equation.	$5 \cdot y + 1 = \left(1 + \sqrt{3} \cdot \sqrt{y}\right)^2$
② Expand expression.	$5 \cdot y + 1 = 1 + 2 \cdot \sqrt{3} \cdot \sqrt{y} + 3 \cdot y$
③ Isolate expression $\sqrt{y}$ in equation.	$\sqrt{y} = \dfrac{y}{\sqrt{3}}$
④ Square both sides of equation.	$y = \dfrac{1}{3} \cdot y^2$
⑤ Subtract expression from both sides of equation.	$0 = \dfrac{1}{3} \cdot y^2 - y$
⑥ Factor expression.	$0 = \dfrac{1}{3} \cdot y \cdot (y - 3)$
⑦ Solve equation.	$0 = \dfrac{1}{3} \cdot y \cdot (y - 3)$ if y is in $\{y1, y2\}$ where $\boxed{y1 = 0}$ $\boxed{y2 = 3}$

Fig. 1.18. User-computer dialogue when using the step-by-step transformation capability of a symbolic-manipulation program to solve the radical equation

$$\sqrt{5y + 1} = 1 + \sqrt{3y}$$

A computer-algebra program could also be used to generate additional symbolic and graphical representations of the solution to the equation

$$\sqrt{5y + 1} = 1 + \sqrt{3y}.$$

As shown in figure 1.19, the values of 0 and 3 can be obtained with a direct-solve command. Graphically, the values $y = 0$ and $y = 3$ are the y-values of the points of intersection of the functions

$$f(y) = \sqrt{5y + 1} \quad \text{and} \quad g(y) = 1 + \sqrt{3y}.$$

Try This: Have students use a symbolic-manipulation program to continue exploring the coefficients of extended products of the form (x − 1)(x − 2)(x − 3) ... (x − k).
Ask them to figure out a rule for the coefficients of the terms of fourth-highest degree, starting with the product (x − 1)(x − 2) • (x − 3)(x − 4).

Some symbolic-manipulation programs can display coefficients for polynomials of the form (x − a)(x − b)(x − c) • (x − d).

If the available symbolic-manipulation program has this capacity, have students consider the expanded form of such polynomials as (x − a)(x − b)(x − c)(x − d) and compare the pattern of coefficients to the coefficients generated for specific values of a, b, c, and so on.

Assessment Matters: This exploration would be a good group-assessment activity following the earlier work in class. Focus not only on whether students can arrive at a correct answer but also on the kinds of interaction they engage in on the way, such as unproductive paths and good insights.

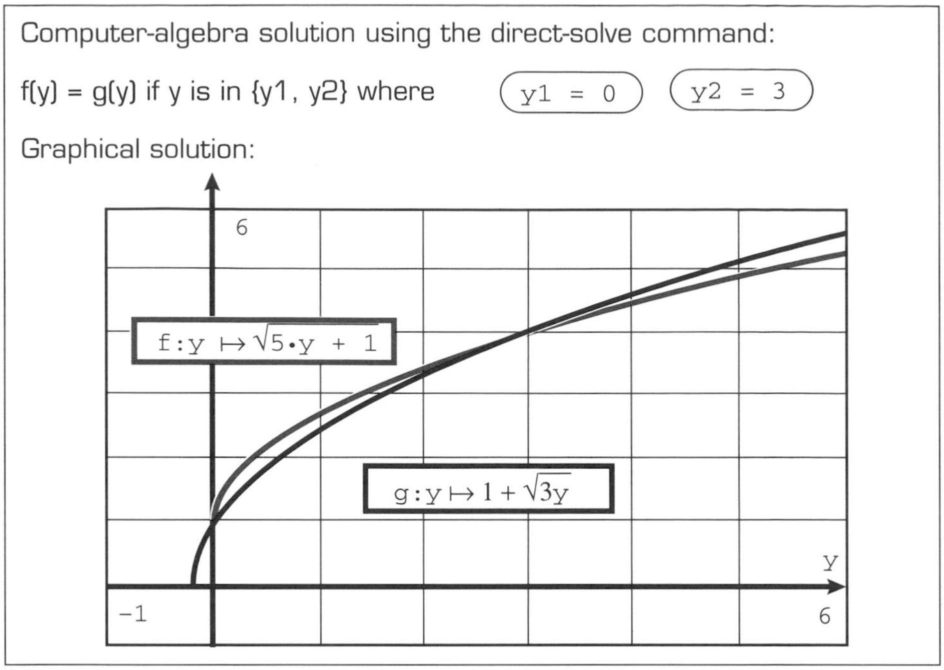

Computer-algebra solution using the direct-solve command:

$f(y) = g(y)$ if y is in {y1, y2} where ⬭ y1 = 0 ⬭ y2 = 3

Graphical solution:

Fig. 1.19. Symbolic and graphical solutions to
$$\sqrt{5y + 1} = 1 + \sqrt{3y}$$

In addition to allowing students to focus on the more global procedures and principles, other reasons exist for this use of a symbolic-manipulation program. For example, sometimes symbolic-manipulation programs will not be able to perform an indicated transformation on a particular form of an equation or expression. The user, therefore, must be familiar with ways to produce other more tractable forms of the expressions.

Symbolic-manipulation tools allow exploratory searches for symbolic patterns.

Since symbolic-manipulation programs make producing symbolic results so accessible, their use in algebra classes opens new opportunities to explore patterns in symbolic expressions. In the next example, a symbolic-manipulation program was used to produce the products so attention could be devoted to looking for patterns in the symbols. It turns out that a rich pattern can be found in the coefficients of the expansions of products of the form (x + 1)(x + 2)(x + 3) ... (x + k).

These patterns can be seen in the following expansions, which were obtained with a symbolic-manipulation program.

$$(x + 1)(x + 2) = x^2 + 3x + 2$$
$$(x + 1)(x + 2)(x + 3) = x^3 + 6x^2 + 11x + 6$$
$$(x + 1)(x + 2)(x + 3)(x + 4) = x^4 + 10x^3 + 35x^2 + 50x + 24$$
$$(x + 1)(x + 2)(x + 3)(x + 4)(x +5) = x^5 + 15x^4 + 85x^3 + 225x^2 + 274x + 120$$
$$(x + 1)(x + 2)(x + 3)(x + 4)(x + 5)(x + 6) = x^6 + 21x^5 + 175x^4 + 735x^3 + 1624x^2 + 1764x + 720$$

The coefficients of the terms of second-highest degree (3, 6, 10, 15, 21, ...) may seem familiar. They are the triangular numbers, the sum of the first k integers.

$$3 = 1 + 2$$
$$6 = 1 + 2 + 3$$
$$10 = 1 + 2 + 3 + 4$$
$$15 = 1 + 2 + 3 + 4 + 5$$
$$21 = 1 + 2 + 3 + 4 + 5 + 6$$

The constant terms may also be familiar, since they are the factorial numbers: 2, 6, 24, 120, 720, and so on.

$$2 = 2!$$
$$6 = 3!$$
$$24 = 4!$$
$$120 = 5!$$
$$720 = 6!$$

The terms of third-highest degree in each polynomial of four or more terms may not be as familiar. After careful analysis, however, we observe that they are the sums of the pairwise products of the first k natural numbers, as illustrated below:

$$11 = (1)(2) + (1)(3) + (2)(3)$$

$$35 = (1)(2) + (1)(3) + (1)(4) + (2)(3) + (2)(4) + (3)(4)$$

$$85 = (1)(2) + (1)(3) + (1)(4) + (1)(5) + (2)(3) + (2)(4) \\ + (2)(5) + (3)(4) + (3)(5) + (4)(5)$$

$$175 = (1)(2) + (1)(3) + (1)(4) + (1)(5) + (1)(6) + (2)(3) + (2)(4) + (2)(5) \\ + (2)(6) + (3)(4) + (3)(5) + (3)(6) + (4)(5) + (4)(6)$$

Spreadsheet Tools

Spreadsheets can create a multirepresentational environment for studying functions. This environment also can be used to analyze functions with techniques previously unavailable to students. Here, a familiar example is analyzed from a spreadsheet point of view.

> ## The Flight of a Baseball
>
> A ball is thrown straight upward with an initial velocity of 120 ft/sec. If the ball was six feet above the ground when it was released, how high did the ball go and how long did it stay in the air?

To prepare for this exploration, have students experiment with and observe their own ball throwing. They can design experiments to find out how fast they can throw a ball and how long they can keep the ball in the air. They can record their results and compare them with the input-output pairs from a given function H for the height of the ball as a function of the time t since the ball was thrown, where

$$H(t) = -16t^2 + 120t + 6.$$

As shown in figure 1.20, students can set up a spreadsheet that will create a table of twenty-six values for H. To create the table of values

Try This: Have students use a spreadsheet to explore the effects of varying the length and width of a rectangle on its area if the perimeter remains the same. Have them explore the effects on the perimeter of keeping the area the same but varying the length and width. Try analogous problems for different types of figures.

Teaching Matters: The iterative definition of functions is exceedingly important when using spreadsheets and similar numerical software. Conversely, in a technological world iteration becomes a powerful way to describe and analyze functional relationships. Compare the linear function $y_{n+1} = y_n + 2$ with $y_0 = 1$, which means to take the present value and add, with the exponential function $y_{n+1} = 2 \cdot y_n$ with $y_0 = 1$, which means to take the present value and multiply. This difference points out very clearly the important relationships of linear models as additive phenomena and exponential models as multiplicative phenomena. Instead of focusing on the functional values, the notation of iteration gives an early emphasis to denoting how the functional values change. Bringing the mathematical concept of change into the early experience of students will give them a powerful way to view their world.

low_t	high_t	inc
0.0	10.00	0.40

t	H(t)	Differences
0.0	6.00	50.56
0.4	51.44	45.44
0.8	91.76	40.32
1.2	126.96	35.20
1.6	157.04	30.80
2.0	182.00	24.96
2.4	201.84	19.84
2.8	216.56	14.72
3.2	226.16	9.60
3.6	230.64	4.48
4.0	230.00	–0.64
4.4	224.24	–5.76
4.8	213.36	–10.88
5.2	197.36	–16.00
5.6	176.24	–21.12
6.0	150.00	–26.24
6.4	118.64	–31.36
6.8	82.16	–36.48
7.2	40.56	–41.60
7.6	–6.16	–46.72
8.0	–58.00	–51.84
8.4	–114.96	–56.96
8.8	–177.04	–62.08
9.2	–244.24	–67.20
9.6	–316.56	–72.32
10.0	–394.00	–77.44

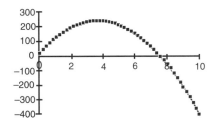

Fig. 1.20. Spreadsheet showing a table of values for H(t) = –16t² + 120t + 6 and the corresponding graph produced with the spreadsheet program Excel (Microsoft)

suggested in the baseball problem, students will need to enter the value for the least value of t, low_t, and the largest value of t, high_t. By letting inc = (high_t – low_t)/25, they can easily create a sequence of twenty-six values of t. Using the graphing capability of spreadsheets, students can next create a graph of H.

Once students have completed this spreadsheet, ask questions like the following:

♦ How can the spreadsheet be used to tell when the ball is going up?

♦ Suggest to students that they add a third column to their sheet that will contain the difference between the value of H for the current value of t and the preceding values of t. Ask them what it means if the differences are positive or negative. [The ball is going up or going down, respectively.] What is special about the values of t where the differences go from positive to negative, and vice versa?

♦ Suppose the differences go from positive to negative between t_1 and t_2. For example, in figure 1.20, the differences go from 4.48 to –0.64 for t-values of 3.6 and 4. What will happen to the table of values if it is reconstructed using t_1 as the lowest input value and t_2 as the highest output value?

Computer-Algebra Systems and Multiply Linked Representations

Computer-algebra systems most often contain symbolic-manipulation programs, graphing programs, and some table or spreadsheet capacity. An important feature of these systems is that the results from one part of the system can be communicated to another part. A variety of computer-algebra systems are now accessible to the computers commonly found in mathematics classrooms at the secondary level. Important features of these systems are their speed, their flexibility, and their ease of use. Computer-algebra systems useful for school algebra programs are listed in the annotated list of computer resources.

These tools allow moving back and forth among graphical, numerical, and symbolic representations.

Computer-algebra systems are perhaps the most flexible tools for moving back and forth among graphical, numerical, and symbolic representations of functions. Using computer-algebra systems, students can freely explore functions of one or two variables, calling on graphical, numerical, and symbolic representations as needed. Although some spreadsheets include numerical and graphical capacity and some symbolic-manipulation programs can be configured to produce tables of function values, the simultaneous and connected access to all three types of representations makes computer-algebra systems so powerful.

Whereas computer-algebra systems allow students to view functions as tables, graphs, or symbolic rules, most such systems require that a function be represented symbolically before it can be transformed. Some programs have been designed to translate a change in the graphical or symbolic representation of a function immediately into the corresponding change in another selected representation. For example, using Function Probe, students "grab" a function graph and perform one or more of the selected range of transformations. The transformation can be a horizontal or vertical translation, a reflection over the horizontal or vertical axis, a reflection over a diagonal, or a stretch in the vertical or horizontal direction. The student can immediately see the effect of the transformation on the function rule and on a related table of values. Figures 1.21, 1.22, and 1.23 illustrate several such sequences from Function Probe. In figure 1.21a, the student has specified a function rule $g1(x) = x^4 + x^3 - 5x^2$, generated a graph for that rule, and created a table of values corresponding to the rule by specifying the starting value, the ending value, and the increment. Figure 1.21b shows the result when the student "grabs" the original graph and drags it up five units. The program then displays the function rule of the resulting graph, $g2(x) = x^4 + x^3 - 5x^2 + 5$, and creates tables of values for the two function rules. Figure 1.21c shows the effect on the graph and table of values of the student's "grabbing" the graph of $y = g2(x)$ and shifting it one unit to the right. The function-rule program gives the transformed rule, $g3(x) = (x - 1)^4 + (x - 1)^3 - 5(x - 1)^2 + 5$.

(a)

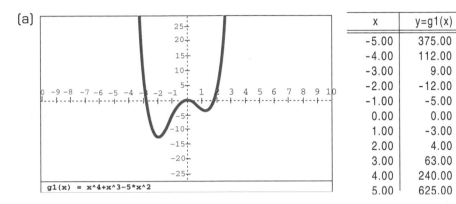

x	y=g1(x)
-5.00	375.00
-4.00	112.00
-3.00	9.00
-2.00	-12.00
-1.00	-5.00
0.00	0.00
1.00	-3.00
2.00	4.00
3.00	63.00
4.00	240.00
5.00	625.00

g1(x) = x^4+x^3-5*x^2

(b)

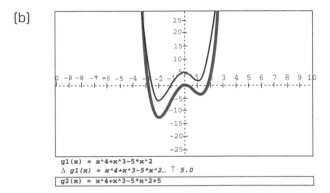

x	y=g1(x)	z=g2(x)
-5.00	375.00	380.00
-4.00	112.00	117.00
-3.00	9.00	14.00
-2.00	-12.00	-7.00
-1.00	-5.00	0.00
0.00	0.00	5.00
1.00	-3.00	2.00
2.00	4.00	9.00
3.00	63.00	68.00
4.00	240.00	245.00
5.00	625.00	630.00

g1(x) = x^4+x^3-5*x^2
Δ g1(x) = x^4+x^3-5*x^2… ↑ 5.0
g2(x) = x^4+x^3-5*x^2+5

(c)

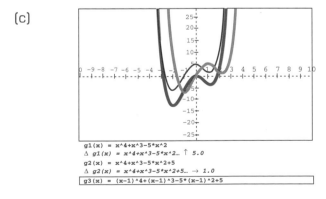

x	y=g1(x)	z=g2(x)	w=g3(x)
-5.00	375.00	380.00	905.00
-4.00	112.00	117.00	380.00
-3.00	9.00	14.00	117.00
-2.00	-12.00	-7.00	14.00
-1.00	-5.00	0.00	-7.00
0.00	0.00	5.00	0.00
1.00	-3.00	2.00	5.00
2.00	4.00	9.00	2.00
3.00	63.00	68.00	9.00
4.00	240.00	245.00	68.00
5.00	625.00	630.00	245.00

g1(x) = x^4+x^3-5*x^2
Δ g1(x) = x^4+x^3-5*x^2… ↑ 5.0
g2(x) = x^4+x^3-5*x^2+5
Δ g2(x) = x^4+x^3-5*x^2+5… → 1.0
g3(x) = (x-1)^4+(x-1)^3-5*(x-1)^2+5

Fig. 1.21. Diagram (a) shows a graph and table of values for $g1(x) = x^4 + x^3 - 5x^2$. Diagram (b) illustrates the graphical and tabular results when $g1(x)$ is translated five units upward. In addition, diagram (c) illustrates the graphical and tabular results when $g1(x)$ is translated five units upward and one unit to the right.

Figure 1.22 shows the graphical, numerical, and symbolic results when the original function, *g*1, is reflected first over the horizontal axis and then over the vertical axis. The simultaneous display of the graphs and tables gives students the opportunity to make observations and conjectures about the effect of these reflections on output values. Finally, figure 1.23 gives the original function and its inverse. The equal horizontal and vertical scales highlight the striking symmetry of the graph and its inverse.

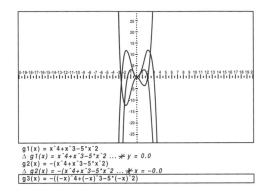

x	y=g1(x)	z=g2(x)	w=g3(x)
-5.00	375.00	-375.00	-625.00
-4.00	112.00	-112.00	-240.00
-3.00	9.00	-9.00	-63.00
-2.00	-12.00	12.00	-4.00
-1.00	-5.00	5.00	3.00
0.00	0.00	-0.00	-0.00
1.00	-3.00	3.00	5.00
2.00	4.00	-4.00	12.00
3.00	63.00	-63.00	-9.00
4.00	240.00	-240.00	-112.00
5.00	625.00	-625.00	-375.00

g1(x) = x`4+x`3–5*x`2
Δ g1(x) = x`4+x`3–5*x`2 ... *y = 0.0
g2(x) = –(x`4+x`3–5*x`2)
Δ g2(x) = –(x`4+x`3–5*x`2 ... *x = –0.0
g3(x) = –((–x)`4+(–x)`3–5*(–x)`2)

Fig. 1.22. The diagram illustrates the graphical and tabular results when g1(x) = x⁴ + x³ – 5x² is reflected over the horizontal axis and then over the vertical axis.

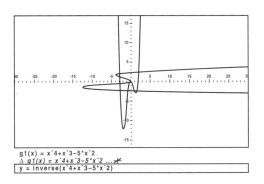

g1(x) = x`4+x`3–5*x`2
Δ g1(x) = x`4+x`3–5*x`2 ... *
y = inverse(x`4+x`3–5*x`2)

Fig. 1.23. The diagram illustrates the graphical result when g1(x) = x⁴ + x³ – 5x² is reflected over the line y = x.

General tools like computer-algebra systems and more specific tools like Function Probe allow users to go back and forth among the traditional graphical, numerical, and symbolic representations of functions. In addition, software is available that generates different types of graphical representations that offer insight into the nature of different types of functions. One example is the Function Family Register, a program designed to help students investigate invariant properties of families of functions. It displays simultaneous graphs of functions and graphical representations of their parameters. Students could study the behavior of a linear function, for example, by choosing the family *U*x + *V*, where *U* and *V* are parameters. The student enters values for the parameters and then sees two different representations of the parameters. For an input of *U* = 1 and *V* = 2, the program would display screens like those shown in figure 1.24. Here (*V*, *U*) values are displayed on the grid.

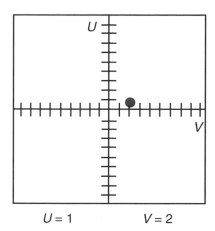 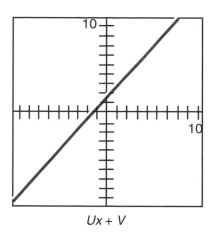

$U = 1$ $V = 2$ $Ux + V$

Fig. 1.24. Output from the Function Family Register program for one function belonging to the family Ux + V. The function displayed is 1x + 2 with a U-value of 1 and a V-value of 2. The left grid displays (V, U) pairs and the right grid displays (x, y) pairs.

Exploring families of functions can take a variety of forms. For example, one exploration might look at the family whose (V, U) pairs form a line of slope –1 such that $V + U = 5$. Figure 1.25 shows a number of graphs of functions belonging to this family. All these graphs intersect in the point (1, 5). Follow-up explorations would ask such questions as "Will all functions in this family contain the point (1, 5)?"

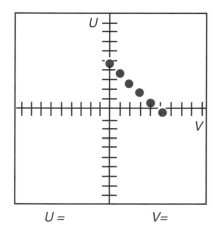 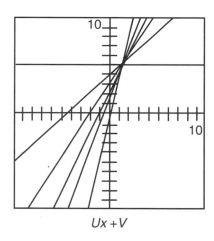

$U =$ $V =$ $Ux + V$

Fig. 1.25. Output from the Function Family Register program for the family Ux + V, with V + U = 5. The (V, U) values for these family numbers are (0, 5), (1, 4), (2, 3), (3, 2), (4, 1), and (5, 0).

In the preceding question, we started with a pattern of (V, U) values and sought characteristics of the lines that generated those (V, U) pairs. Another type of question we might pursue is to find the pattern of (V, U) pairs that are associated with a given set of lines. For example, we might explore the question, "What pattern of (V, U) pairs might be associated with parallel lines of the form $Ux +V$?" If the lines are parallel, then the U-values are the same. The set of parallel lines, therefore, will generate (V, U) pairs that have the same U-value but varying V-values. That is, the (V, U) values lie on a single horizontal line. We could follow up with this question: "What characterizes the functions that generate the (V, U) values that lie on a single vertical line?" We can use the Function Family Register to explore a large variety of other questions about linear and quadratic functions and their parameter representations.

Myriad other explorations could be conducted with linear, quadratic, exponential, and rational functions. The common question for each such exploration is, "What are the invariants?" In the example shown in figure 1.25, the lines all intersected at the point (1, 5). The property of "containing the points (1, 5)" was invariant.

This genre of computer software contains a wide range of programs with a considerable variety of different representations for functions. Many of these programs are described in the annotated list of computer resources.

Exploratory and Catalyst Software

Exploratory software encourages students to explore mathematical relationships through games. One of the earliest and most popular of this genre is Green Globs. In this program, a series of "globs" are placed at random lattice points on the screen. Students must write equations for curves that pass through the greatest possible number of "globs." In one version, "shot absorbers" designate points through which the curve cannot pass. A sample of a game in progress is shown in figure 1.26.

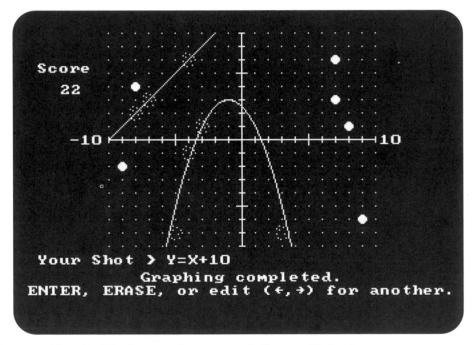

Fig. 1.26. A sample game of Green Globs in progress

Video Technology and Simulations

Video technology opens the real world to mathematical exploration by secondary students. Even a videocassette player can bring the real world into the classroom for mathematical analysis. When a video recording is combined with other technology, data can be recorded from real-life events and analyzed with algebraic computing tools. One intriguing example of the potential for such technology comes from Andee Rubin. Students were presented the following problem and asked to collect and analyze data to support their conclusions.

Why does one person run faster than another?

To find an answer, students measured arm lengths, leg lengths, and strides and ran and videotaped races.

They collected additional data from the videotapes and used computer software to analyze the data by looking at patterns in tables and graphs. Their data supported a conjecture that some of the speed was attributable to stride length. The importance of activities like these is that the data and the analysis were generated by students in ways that made sense to them.

The whole issue of what gives one competitor an edge over another in a sporting event has benefited greatly from video technology. Sports physiologists use videos of Olympic training sessions to study the relationships among body angles, lengths of limbs, and styles and the success of a particular athlete.

In addition to allowing access to real-world events, video technology furnishes new opportunities for students to use algebraic concepts and tools to analyze simulations of real-world events. With Interactive Physics II, for example, students can run experiments and collect data about relationships among variables. They can then use their knowledge of algebraic functions to generate rules to describe the relationships.

The screen dumps from Interactive Physics II in figure 1.27 show sequential attempts at determining the initial speed required to make a soccer ball travel three meters in the air. Students can run the simulation numerous times, collecting pairs of data on the time in the air versus the initial speed. They can then apply their knowledge of algebraic relationships along with appropriate computing tools to generate a rule describing the relationship between the initial speed and the distance traveled before the ball hits the ground. Although this particular simulation is part of the Interactive Physics II package, students can use the package to create and run their own experiments.

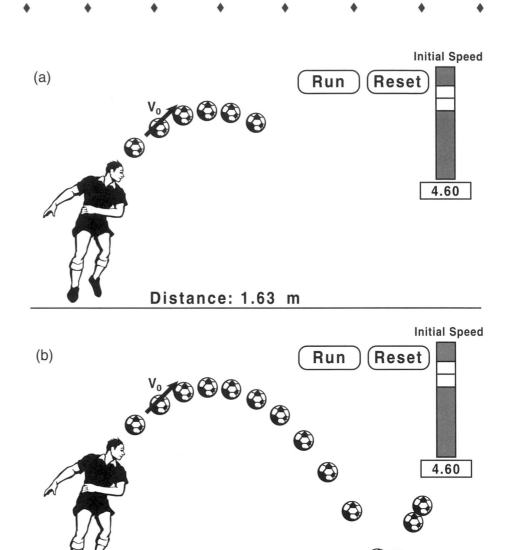

(a)

Initial Speed

Run Reset

4.60

Distance: 1.63 m

(b)

Initial Speed

Run Reset

4.60

Distance: 3.85 m

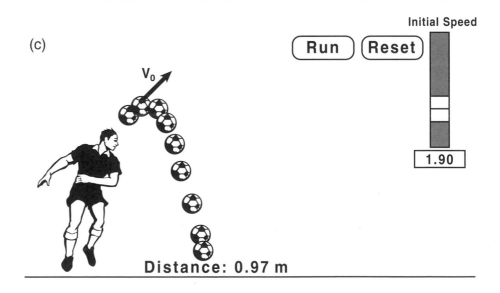

(c)

Initial Speed

Run Reset

1.90

Distance: 0.97 m

Fig. 1.27. Diagrams (a) and (b) show the beginning of the path of a soccer ball when the initial speed is 4.60 meters a second. Diagram (c) shows the path of a soccer ball when the initial speed is 1.90 meters a second.

WHAT IS THE FUTURE CONTENT OF ALGEBRA IN A TECHNOLOGICAL WORLD?

Because of the importance of algebra in describing and understanding the world, because of the increasing access that students have to graphical, numerical, and symbolic computing technology, and because of the newfound commitment of mathematics teachers to building a strong foundation in mathematics for *every* student, school algebra of the future is likely to rely on a vastly different body of content than it does currently. The content of school algebra will be transformed through changes driven by a variety of perspectives:

◆ Changes in organization

◆ Changes in sequencing

◆ Changes in emphasis

◆ Changes in focus

◆ Changes in representation worlds

Changes in Organization

In a technological world, algebra is a way of thinking rather than merely a set of techniques for completing a predesignated set of tasks. Algebraic thinking is a fundamental strand within the fabric of mathematics, and major ideas can be developed through targeted algebra courses or through strands in an integrated curriculum. Either way, algebraic thinking should permeate the school mathematics curriculum from kindergarten through grade 12 and algebraic ideas should be applied in every school mathematics course. Algebraic ideas are useful in data analysis as students explore functional relationships in the data they gather, in discrete mathematics as students analyze iterative situations, and in geometry as students quantify and analyze the dynamic relationships among characteristics of figures.

Changes in Sequencing: Skills and Concepts

With computers and calculators available for producing graphs, tables, and symbolic manipulation, refining by-hand routine skills no longer must precede developing broader conceptual ideas. New skills and concepts are needed in a technological world. It is not as important for students to know how to sketch a graph by hand as it is for them to understand what the graph means and how it can be used to enhance their understanding of the function or relation the graph is describing. Whereas the ability to factor was a necessary tool in the pretechnological algebraic classroom, the ability to understand a factor is much more important in the technological algebraic world. With the vast array of graphical, numerical, and symbolic tools now in classrooms (or waiting patiently at the doors of some mathematics classrooms), routine manipulative skills and procedures will no longer dominate the content of school algebra. Instead, the algebra that finds its way into the school mathematics classrooms of the future will begin with, and concentrate on, concepts and applications instead of routine skills.

Changes in Emphasis: Less Complex Manipulation by Hand

In the past, students experienced algebra by practicing each skill or procedure in some of its more complicated forms. Teachers tested whether students really understood basic algebra principles by asking them to perform exercises that required them to manipulate some of the more complex expressions and equations. Computing technology compels us to rethink this position.

In the near future, all but the most basic symbolic manipulation will probably be accomplished with computing tools. The intelligent user, however, will know what to ask from the computing tools and how to interpret the results they generate. Just as scientific and four-function calculators helped refocus the teaching of arithmetic on number sense, computing technology (graphical, symbolic, and numerical) can help refocus the teaching of algebra on the development of students' symbol sense.

At first, teachers and students may find it difficult to transform their thinking so that symbolic-manipulation skills are no longer the central focus of school algebra. Even in a technological world, students may want to do some algebraic manipulation by hand. But that manipulation will be of the most basic form, and its mastery must not be the focus of school algebra encounters. Learning these very basic manipulation skills is likely to come easily within the context of exploring bigger ideas and within encounters with mathematical modeling and problem solving.

Changes in Focus: Emphasis on Doing Mathematics instead of Redoing It

In the spirit of the NCTM's *Curriculum and Evaluation Standards for School Mathematics* (1989), algebra in a technological world can make students active partners in the exploration of mathematical ideas. They can use the full power of computing tools to investigate the real and mathematical worlds around them.

At Forrestdale School in Rumson, New Jersey, a group of seventh and eighth graders, enrolled in computing-intensive algebra classes whose focus was mathematical modeling and functions instead of symbolic manipulations, engaged in a discussion of what they thought about this kind of algebra. Among other things, they pointed out that *this kind of algebra emphasizes doing mathematics instead of redoing it.*

Mathematical modeling

Mathematical modeling takes a center-stage role in technologically rich mathematics courses. Without technological approaches, algebra students of the past commonly encountered real-world applications that were contrived or overly simplified. Technology opens new worlds of mathematical modeling to secondary school students, making it possible for them to experience more-realistic problems without spending all their time mastering the related symbolic-manipulation skills. This newfound role for modeling, however, generates new obligations to deal with the limitations of particular models and with whether particular inferences can be drawn from these models.

An important difference exists between current traditional algebra courses and those that capitalize on technology to explore mathematical models. Present-day high school algebra experiences are like the experiences of a Little Leaguer who practices catching and hitting but never plays the game. Technologically rich mathematics classrooms invite students to play the real game.

Function families

Today's algebra experiences confine students to studying a very limited set of functions; commonly, students study only the behavior of linear and quadratic functions. Their experiences leave many with the misconception that functions are not really useful for describing their world. With technology, instead of refining their knowledge of a single family (or two) of functions, students have the opportunity to look at functions from a broader perspective. They become aware of essential properties of a variety of families of functions, and they progress toward analyzing properties of combinations and compositions of those families.

Changes in Representation Worlds

Probably the most fundamental shift in the future of school algebra is the new role for representation worlds. Two aspects of technology are centrally influential in that shift: first, a greater variety of types of mathematical representations are available to students; and second, technology links those representations in an impressive range of ways.

With the advent of the graphing calculator, graphical representations have taken the mathematics classroom by storm. Access to graphical representations have invited students to explore new mathematical worlds and to wrest new meaning from familiar mathematical worlds. Visual representations have opened mathematical ideas to a previously uninterested audience.

Along with graphical representations, technology has made both numerical and symbolic representations much more convenient. Students at all levels can view functions from all three perspectives, and new software gives students interactive links among representations.

Henry Pollak (1987) constructed a list of expectations for the mathematically literate worker who will enter the workforce at the start of the twenty-first century. This list, referred to in the *Curriculum and Evaluation Standards* (NCTM 1989, p. 4), reflects quite well a new direction for algebra and for school mathematics in general. Workers will need to have—

♦ the ability to set up problems with the appropriate operations;

♦ knowledge of a variety of techniques to approach and work on problems;

♦ an understanding of the underlying mathematical features of a problem;

♦ the ability to work with others on problems;

♦ the ability to see the applicability of mathematical ideas to common and complex problems;

♦ preparation for open problem situations, since most real problems are not well formulated;

♦ a belief in the utility and value of mathematics.

Technologically rich algebra experiences, in which students work together to model and investigate real-world and mathematical situations, have almost boundless potential for addressing this challenge.

The following chapters will address how the increasingly available technology and the consequent new mathematical opportunities will affect the teaching, learning, and content of algebra in the future technological world. Chapter 2 addresses differences in teaching and learning and their effects on assessment; chapters 3 and 4 characterize a functions approach to algebra; chapter 5 addresses the role of matrices in algebra; and chapter 6 tackles the issues of the changing face of symbolic manipulation and the new role of symbol sense in technologically rich algebra classrooms.

ACTIVITY 1

THE EFFECTS OF MONOMIAL TERMS ON POLYNOMIAL FUNCTIONS

Each of the following graphs is the graph of a polynomial function of degree less than 6 and with integer coefficients between –5 and 5. Use a computer or calculator function grapher to find a polynomial function rule to match each graph.

Record all the function rules tried as well as the final answer. Explain how the trial function rules were used to arrive at the final answer.

1.

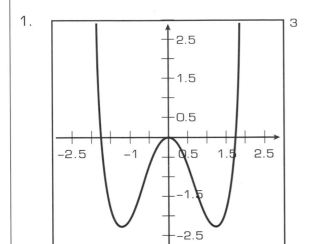

Trial Function Rules

Trial function rule	General shape of graph

Final function rule: _____

Reasoning: _____

2.

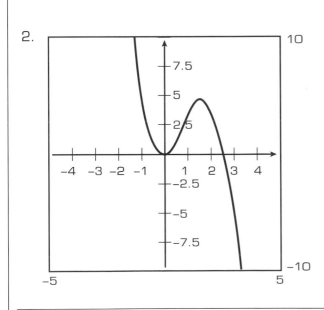

Trial Function Rules

Trial function rule	General shape of graph

Final function rule: _____

Reasoning: _____

3.

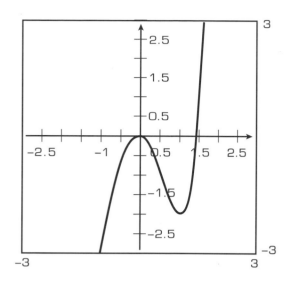

Trial Function Rules

Trial function rule	General shape of graph

Final function rule: _____

Reasoning: _____

_____

_____

4.

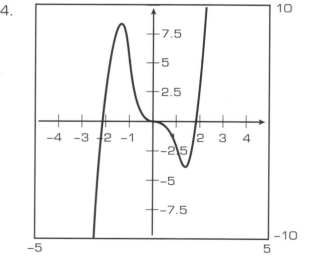

Trial Function Rules

Trial function rule	General shape of graph

Final function rule: _____

Reasoning: _____

_____

_____

ACTIVITY 2
OIL AND WATER DON'T MIX

The United States imports nearly 200 000 000 gallons of oil every day of the year. Unfortunately, some of that oil gets spilled on the way to its destination to be refined into fuel for cars and homes. When a major oil spill occurs on an ocean or river or lake, the shippers try to recover as much oil as possible. The recovery is possible because most oil floats to the surface of the water and forms an oil slick.

Do you know the properties of oil and other materials that determine whether they will float or sink in water?

One part of the explanation for "floaters" and "sinkers" is a property of all matter called *density*. The density of oil or water, or of any other substances, can be measured with some simple experiments and a linear model of data.

Getting Ready: To compare the density of oil and water, the research group will need some common science equipment—a centigram balance and graduated cylinders for measuring the volume of liquid—a supply of water, and some kind of oil, such as cooking oil or automobile oil.

Collecting the Data: Use the graduated cylinder and the balance to find the mass of several different volumes of oil and water. Begin by measuring the mass of the empty cylinder. Then add oil to the cylinder in amounts of two cubic centimeters, measuring the mass after each addition. Report the data in the table below.

Volume of oil in cc	0	2	4	6	8	10	12	14	16
Mass in grams									

Repeat the measurement of mass for various amounts of water. Record the data in the table below.

Volume of water in cc	0	2	4	6	8	10	12	14	16
Mass in grams									

Analyzing the Data: Make a scatterplot for each data set to search for possible patterns. If the patterns are approximately linear, find the median-fit line and its equation for each data set.

Compare and contrast the patterns in the tables, graphs, and equations.

♦ How are the (*volume, mass*) patterns for oil and water most similar and how are they different?

♦ What do those patterns tell about the properties of the two liquids?

In the linear models relating volume and mass for oil and water, the slopes of the graphs give the densities of the two liquids.

1. What are the slopes of the linear models found for (*volume, mass*) data patterns for oil and water? What do those numbers tell about the meaning of *density*?

2. How do the differences in the densities of oil and water help explain oil slicks formed by oil floating on the top of water?

3. If linear models like $y = b + ax$ are found for the (*volume, mass*) data patterns, what does the value of a tell in each case? How could data about density be produced that would give a linear model of the simplest form, $y = ax$?

4. Use the models that relate mass and volume to write and solve equations and inequalities matching each of these questions:

   (*a*) What is the mass of 50 cubic centimeters of oil?

   (*b*) How much oil would give a mass of 250 grams?

   (*c*) What volumes of oil would give a mass of at least 500 grams?

5. A wooden block usually will float in water. However, a lead weight dropped into water will not float.

   (*a*) What does floating or sinking tell about the density of a material in relation to plain water?

   (*b*) How would graphs of (*volume, mass*) data for lead or wood compare with those of oil and water?

6. A can of Diet Coke dropped into a pan of warm water will generally float. A can of regular Coke will generally sink. Diet Coke poured out of the can into the water will not form a "Coke slick" on the surface of the water. What does this experiment tell about factors other than density that are involved in the formation of an oil slick?

Adapted from the "Linear Models" unit, *Core-Plus Mathematics Project.*

CHAPTER 2
CHANGES IN LEARNING AND THEIR CONSEQUENCES FOR TEACHING AND ASSESSING

Technology enters the mathematics classroom and brings with it changes in the learning environment. Two major questions arise: What changes in the learning of algebra occur when technology enters the classroom? What are the consequences of these changes on teaching and assessing algebraic understanding?

CHANGES IN LEARNING

Student Roles

The most obvious changes are probably the differences in what students actually do in the classroom. Students studying algebra in a technological classroom can ask their own questions and pose their own problems, just as they could in a traditional algebra classroom. However, technology allows students to take a more active role in their own learning. Computing tools offer new avenues by which students can try to solve the problems they pose.

Students who are aware that temperature in degrees Fahrenheit is a function of the temperature in degrees Celsius according to the rule $F(C) = (9/5)C + 32$ may hear that the temperature today is 78°F and ask themselves, "What is today's temperature in degrees Celsius?" Traditionally, students first had to master the symbolic-manipulation skills to solve the equation $78 = (9/5)C + 32$. Given technology, students have available multiple methods for solving their problem. For example, they may choose to produce successive refinements of tables of values, such as those shown in figure 2.1; they read an approximate answer, 25.5°C. This ability to pursue different problems of their own design forces students to rely less on their teachers and more on themselves and their peers to assess the success of their small-group or individual work.

C	$F(C)$		C	$F(C)$		C	$F(C)$
0	32		20	68		24	75.2
20	68	← 78	24	75.2	← 78	25	77
40	104		28	82.4		26	78.8
60	140		32	89.6		27	80.6
80	176		36	96.8		28	82.4
100	212		40	104		29	84.2

Fig. 2.1. Tables of values to solve the equation 78 = (9/5)C + 32

Because technological tools open so many options, students need to become intelligent consumers in terms of which tools and which options of those tools they choose to use. For example, suppose students want to find the radius of a sphere for which the surface area and the volume are of equal magnitude. The students may try to solve the system of equations

$$4\pi r^2 = k$$

$$\frac{4}{3}\pi r^3 = k,$$

where r is the unknown radius and k is the common value of the surface area and volume. Many tools currently available (for example, the Mathematics Exploration Toolkit) when asked for a solution to this system respond with either an error message or a comment such as "no solutions found." Students should recognize the need to consider a different symbolic-solution path, such as solving $4\pi r^2 = (4/3)\pi r^3$, or to try such alternative methods as determining the point of intersection of the curves given by $y = 4\pi r^2$ and $y = (4/3)\pi r^3$ or constructing tables of values and searching for similar output values. Figure 2.2 illustrates the latter two alternative methods.

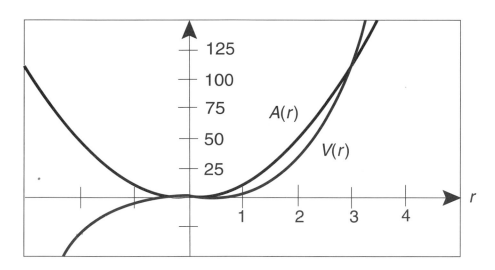

Radius	Surface Area	Volume
0	0.00000	0.00000
1	12.566	4.1888
2	50.265	33.51
3	113.1	113.1
4	201.06	268.08
5	314.16	523.6

Fig. 2.2. Two alternative methods for solving systems of equations

Another implication of this freedom to pose and pursue original problems is the need for students to be comfortable with their mathematical freedom. No longer is there a single acceptable solution process and often no longer a single correct answer. Graphs and tables are equally acceptable tools for answering questions such as those of the surface-area-and-volume problem. Also, students need to contend with temporary or extended uncertainty. For example, when the students attempted to determine the radius value, they did not necessarily know that their tool could not directly solve a nonlinear system or that a table would help them arrive very quickly at 0 and 3, the exact solutions to the equation. The students had to persevere and to evaluate their progress—or lack of it. The students' role in learning expands in the technological environment to include management and monitoring tasks that had traditionally been one of the teacher's primary responsibilities.

Mathematical Thought

Students learning algebra in a technological environment should not only be doing different things; they should also be thinking different things. For example, students should be asking "what if" questions and then planning ways to follow up on those questions. Consider the students who have used technology-based direct-solve commands to solve linear equations and who now are using the direct-solve commands to solve quadratic equations. These students see that the tool produces two solutions for most quadratic equations, whereas it gave only one solution for linear equations. These learners may now ask such questions as "What if we change the coefficients; when will we get two real-number solutions?" or "What if we increase the degree of the equation; will a cubic equation have more roots than a quadratic equation?"

Students should also be thinking about how they can verify or refute what they believe is mathematically true and how they can convince themselves and others of the truth or error of these proposed ideas. Sometimes this verification may be based on technology–generated examples and counterexamples. The students, in trying to explain why a quadratic equation may have no more than two roots, might try graphing quadratic equations of the form $y = ax^2 + bx + c$ and build an argument on the basis of the shape of the graphs. Their computer-generated examples may suggest that the sign of a will determine whether the graph opens upward or downward, but these examples will not supply actual proof; the examples do not include all possible quadratic equations. To justify their conjecture, the students may reason from the general form of the equation to determine the end behavior of the curve as follows (though perhaps not in these words): when the value of $|x|$ is very large, ax^2 is in fact much greater than $|bx + c|$, and so ax^2 "dominates." Then, as x becomes infinitely large, the value of $ax^2 + bx + c$ becomes infinitely large if $a > 0$ or infinitely small if $a < 0$.

At times, a single computer-generated example is sufficient evidence to refute a conjecture. For example, students may contend that the graphs of equations of the form $y = ax^3 + bx^2 + cx + d$ will always have the general shape of the graph of either $y = x^3$ or $y = -x^3$. A graph of $y = x^3 - 3x^2 + 2x$ in a suitable window should quickly convince them that their initial idea was incomplete—the graphs of cubic equations may "turn" as well as "bend." Yet these students should also recognize that the computer-generated examples offer evidence, but not proof, that a cubic equation *must have* a graph of one of the four general shapes shown in figure 2.3.

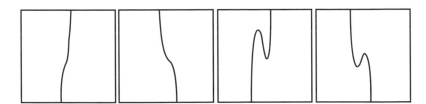

Fig. 2.3. General shapes of graphs of cubic equations

Using technology to form and refine mathematical ideas is not always as clean as the preceding examples may suggest. Students need to question regularly and interpret carefully the evidence they generate. For example, consider the two views of the graph of $y = x^3 - 3x^2 + 2x$ shown in figure 2.4. The students who generated the top graph would benefit from an appropriate skepticism about the image they see. Their interpretation of the graphic image must incorporate the effects of scale on the appearance of the graph. These students should be ready to zoom in or trace the curve in the region near the origin, thus noting that the graph "turns," since the value of y is 0 for the x-values of 1 and 2 as well as when the value of x is 0.

Technology seems not only to allow but to encourage students to ask "what if" questions, to seek answers to these questions, and to generate evidence to support or refute mathematical claims. The major implication of these changes with respect to the role of the learner is monumental: students *themselves* have the *power* to say, "*I* am capable of determining what is mathematical truth."

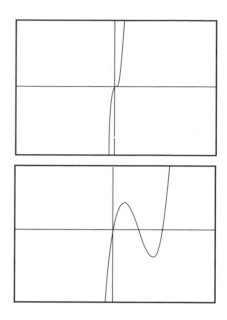

Fig. 2.4. Two views of the graph of $y = x^3 - 3x^2 + 2x$ produced by Graph Wiz software

Readiness Redefined

Technology may aid algebra students in reaching lofty levels of mathematical sophistication. If this sophistication is to be the end product of their algebra-learning experience, what do students need to know before they enter the algebra-learning environment? At least two essential preliminary requirements should be met. Number sense seems to be one aspect of algebra readiness. How could students reason through the end behavior of graphs of equations of the form $y = ax^2 + bx + c$ without some sense of what happens when large numbers are squared? Students also must be able to read and write mathematics as well as voice their mathematical ideas and listen carefully to the ideas of others. How else could classmates understand a student's newly generated "what if" question or debate the extent to which the computer-generated evidence adequately supports a mathematical proposition?

Emphasis now will be on prerequisites that have been of little importance in a traditional algebra class, where writing and reading and speaking and listening were limited to the exchange of predominantly symbolic statements. The essence of what is now needed for students to achieve in secondary school algebra has changed. Previously used measures of algebraic readiness will no longer be sufficient predictors of the extent to which students will succeed in learning and doing algebra in a technological world.

Furthermore, it seems that as schools adopt the NCTM's curriculum standards throughout their K–12 mathematics program, students will no longer "begin" their algebra-learning experiences at the secondary level (NCTM 1989). Youngsters will enter their more formal study of algebraic concepts and skills with a rich mathematical background that includes skills for communicating mathematically, comfort with using multiple representations, acceptance of multiple solution paths, and knowledge of how technological tools are mathematically helpful. Future students will have had early experiences in algebraic reasoning and should be far better prepared than their predecessors to learn later algebraic ideas. This observation suggests the need for continuing not only to redefine algebra readiness but also to reconsider what should be the essence of the high school algebra experience.

Later Mathematical Development

Students who experience the richness of algebra in a technology setting should be better prepared to engage in further study of mathematics for several reasons. First, such concepts as function and extrema values that clearly underlie, or are traditionally new ideas in, advanced mathematics courses are notions with which these students will already be familiar. In addition, these learners have daily experience in relating various representations of the same mathematical phenomena. For example, students who have explored the graphs of linear, quadratic, and cubic functions will have a strong conceptual basis on which to ground such notions as limits that are more powerfully conveyed through a combination of graphic and symbolic modes.

Technology-wise students have varied and numerous opportunities to work with functions and nonsymbolic manipulative methods for solving equations involving these functions. When they encounter exponential, logarithmic, and trigonometric functions—if they have not already!—these learners have well-formed notions about functions as relationships among quantities that change. In addition, these students have powerful numerical and graphical strategies for solving equations that are beyond the limits of usual by-hand manipulation skills. Consider one example of the mathematical advantage of these students. Before graphing the function given by $f(x) = \tan(5x)$, these students can immediately invoke previous experiences and ask relevant questions such as the following, in perhaps less formal terms: "Is the function monotone?" "Does it have a maximum or minimum value?" "Does it have any zeros, and if so, what are they?" "Does it have a vertical intercept?" "How many solutions, if any, would equations like $\tan(5x) = k$ have for any constant k?"

Another advantage is the students' extensive exposure to mathematical structures. For example, families (or classes) of functions are a common topic in technologically strong algebra programs. Function families with function addition can be the basis of a discussion of algebraic structures and properties. Students who use computer-algebra systems develop a repertoire of symbolic and graphical examples and nonexamples of closure, commutativity, and associativity as they observe the results of "simplify" and "graph" commands on the sum, product, difference, or quotient of two or more polynomial function rules. Technology-based activity in learning algebra gives students robust experiences that they may carry into their future study of mathematics at the high school and college level.

CHANGES IN TEACHING

Instruction

Drastic differences abound in what students using technological tools can see, do, and grasp about sophisticated mathematical ideas. In a way, these changes are almost frightening in that they will put much greater demands on the teachers who work with these students. In the classroom, the teacher's role will no longer be that of posing the focal problems, providing all pertinent mathematical information, and directing the solution activities. Technology introduces a setting in which teachers facilitate, rather than direct, problem solving. Furthermore, as students become accustomed to asking questions of their own design, the number of questions they ask and the difficulty of those questions will likely increase.

Teachers will need to orchestrate the activities of the whole class in ways that promote the desired mathematical growth of all students but still allow students to share novel and interesting mathematical insights. With their growth in problem posing comes the inevitable likelihood that students will begin to ask questions that even the most experienced and knowledgeable teachers have not yet pursued. In this way, technology forces teachers to become coinvestigators who are learning at the same time as their students. No longer can teachers function as experts who merely wait for students to travel well-worn paths that lead directly to familiar conclusions. Becoming comfortable in this new role as a fellow learner is a difficult task, cognitively as well as affectively.

Other challenges exist inside and outside the classroom. Teachers cease to be people who deliver the regular daily dose of a prescribed curriculum. Instead, they must respond to the myriad of ideas that students introduce in ways that cannot be explicitly defined before that particular class meeting. The instructional decisions that are made in classrooms will be based more on a clear understanding of the overall goals of the high school mathematics curriculum and less on strict adherence to a single, specific daily objective. In essence, each teacher must capitalize on more-frequent and less predictable teachable moments and thereby develop and implement an algebra curriculum in concert with the understandings and shared experiences of the students in a given class. To maximize the potential that technology and exploration offer, teachers will need to be appropriately fluent in the technology and to have ready access to a variety of physical and experiential resources that enhance a lesson as it unfolds. This change requires constantly planning, critiquing, and replanning daily activities in ways different from those traditionally needed for successful teaching.

Outside the classroom, teachers need to stay abreast of current technology, assemble different sets of activities and stories, and redefine expectations. In the classroom, using technology means altering or not altering the direction of the course on the basis of multiple forces, including the excitement generated by the students' or teacher's technology-based observations and the fervor of students' "what if" questions.

Assessment

The redefinition of expectations goes hand in hand with the need to revise assessment procedures. The evaluation of students' understanding, as well as the evaluation of teaching performance, must undergo changes as profound as those in instruction and curriculum. As individual students or small groups of students pursue their own questions, differences among classmates' understanding of the topic of study will naturally arise. Teachers need to remain aware of those differences and to adjust evaluation to reflect those differences in students' opportunities to learn. Furthermore, the nature of what is tested may need substantial revision. More emphasis will be placed on conceptual understanding than on the execution of algorithms, and more attention will be given to the processes by which mathematical truths are obtained than to the outcomes of those processes. The changing emphases, as well as the changing content, suggest a need for assessment instruments that differ both in content and in form.

Evoking Results. Issues of the content and form of assessment can be illustrated by considering the earlier lesson in which students explored the shapes of the graphs of quadratic and cubic functions. We can argue that the students who explored the graphs of the cubic and quadratic functions really understand the mathematics if they can apply that knowledge to new cubic and quadratic functions. A legitimate evaluation task from this perspective may require students to discuss the potential

Assessment Matters: The various types of questions in this section illustrate an important assessment principle: A good assessment provides the teacher with far more information than just whether the student got a correct answer. These questions give a great deal of information about what the students are thinking, how they view mathematics, and what they do and do not understand.

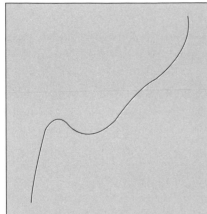

Could the graph shown above be a graph of a quadratic function?

Of a cubic function?

Explain the reasoning behind your answers.

Assessment Matters: The item above could be scored with 0–4 points, with 1 point given for each correct answer and 0–2 points given for the explanation.

graphs as representations of quadratic or cubic functions. The problem at the left gives a potential item that could appear on a paper-and-pencil examination or quiz and that requires the application of student-developed knowledge.

Observing Process. However, implicit in the students' work with the quadratic and cubic functions was the development of a powerful mathematical notion—the process of how they explore a family of functions. This knowledge of exploration includes many mathematically fundamental issues, such as the exhaustion of all possible cases, the impact of an infinitely large set of possible examples, the difference between an example and a counterexample, and the interdependence of companion outcomes in corresponding symbolic and graphical representations. Legitimate assessment activities explore the students' understandings of this exploratory way of developing mathematical truth as well as the factual outcomes to which the exploration led. The task below requires students to attend to the process and not the product of their previous in-class investigations.

This task would be appropriately presented to a pair of students who had access to computing tools. As the students work, the teacher would record, perhaps with the assistance of audio or video equipment, such central characteristics of the students' work as the following:

♦ Functions that the students consider to be examples of quartic functions and the order in which they consider these examples

♦ Forms of the examples they consider and from which they reason, such as graphs and tables

♦ Conjectures the students form on the basis of their examples

♦ What the students say and do to explain their ideas to each other

♦ Reasons or evidence, other than the example functions, that the students use to justify or reformulate their conjectures

When we worked with the graphs of cubic functions in class, we determined that the graph of any cubic function must have one of the four general shapes shown below.

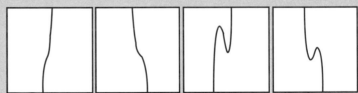

Determine all possible shapes that the graph of any quartic, or fourth degree, function can have.

In analyzing the students' work, the teacher would look more deeply at the characteristics of their interactions with the mathematics as well as at their interactions with the tools and with each other. The following questions may help focus attention on the mathematical power of the students' reasoning:

♦ To what extent are the students' conclusions consistent with the examples?

♦ To what extent do the students use alternative sources of information? That is, do they use tables as well as graphs to identify the end behavior of the function graphs?

♦ To what extent do they choose new examples that are based on their observations about previous examples?

♦ To what extent do they use a set of examples that is indicative of all possible combinations of the signs of the coefficients?

♦ Once they arrive at a conjecture, how, if at all, do they attempt to verify it?

♦ How do the students communicate with each other?

One goal of the function-family task might be to give insight into students' progress and current understanding for the purpose of monitoring instruction. The teacher who observes pairs of students working on this task may note patterns that occur across several pairs. For example, the teacher may notice that all but one pair of students create only graphs and that most pairs never manipulate these graphs, that is, they never change the viewing window or trace the graph. This teacher might act on these observations by asking the students who do use alternative representations or who do manipulate the graphs to share these ideas with their classmates. A whole-class discussion could then focus on the question "What new information, if any, might we learn about quartic functions if we look at other representations or if we look at the same representation in different ways?"

The function-family exploration task can also be used if the assessment goal is student evaluation. In this situation, the teacher's intention may be to gather some indication of the students' accomplishments in terms of communicating mathematically. Figure 2.5 illustrates one possible scoring rubric designed for use with the function-family exploration task given this general category and its corresponding areas of interest.

Assessment Matters: To encourage communication between student-partners, it is reasonable to construct a rubric to assess students' mathematical communications. One possible rubric to assess the communication between students working in pairs on computer explorations is presented in figure 2.5. This rubric makes an important point about assessment. Here, the issue is assessing communication, not the success of the problem solving. Students could achieve 2/3 of the total points through appropriate communication and unsuccessful problem solving. Because of this phenomenon, communications rubrics are most often used in conjunction with a rubric targeted on strictly mathematical issues.

Areas of Interest	Points	Qualities of Work
Communicating mathematically		
Explanations are understood by partners.	2	Consistently seeks to understand and to be understood
	1	Occasionally seeks to understand and to be understood
	0	Lacks sharing *or* regularly ignores partner's ideas and work
Each student responds to his or her partner's ideas.	2	Regularly acts on or clarifies partner's ideas
	1	Occasionally acts on or clarifies partner's ideas
	0	Ignores or demeans partner's ideas
Communications are mathematically correct.		Uses central terms and symbols correctly:
	2	Most of the time
	1	Occasionally
	0	Rarely, if ever

Fig. 2.5. One possible scoring rubric designed for use with the function-family exploration task

Each pair of students would receive a number of points for each of the three "areas of interest." The number of points for any one of these areas would be the maximum number of points possible according to the qualities of the pair's work. For example, two students start with these examples of quartic functions, saying that they are omitting the constant term because "that will probably only make the graphs slide up and down":

$$a(x) = x^4 + 3x^3 + 2x^2 + x$$
$$b(x) = x^4 + 3x^3 + 2x^2 + 5x$$
$$c(x) = x^4 + 3x^3 + 2x^2 - 6x$$

They talk about choosing 1, 5, and –6 as the linear coefficients because they want to see whether the sign and size of the linear coefficient have any graphical effects. Their work produces the graph in figure 2.6. The students note that all three graphs appear to be nearly identical within this viewing window. They then decide to zoom in on the graphs, and they obtain the view in figure 2.7.

They say that changing the value of the linear coefficient seems to cause the graph to move around a little, but the graph of any of their functions still looks like a "flat-bottomed bowl that goes up forever." One student then says that maybe the linear term is not the one about which they should worry. The cubic term may be more important, since "x cubed is much bigger than x or even x squared when the values of x are big." They then create the image in figure 2.8 as they test their next two examples:

$$d(x) = x^4 + 5x^3 + 2x^2 - 6x$$
$$e(x) = x^4 - 8x^3 + 2x^2 - 6x$$

One student says that it looks like the graphs will either be parabolas with "wavy or flat bottoms" or be shaped like the graphs of the cubic functions. The other student asks if it can be like both the cubic and the quadratic. The first student says, "It has both a cubic and a quadratic term in it, so." The second student then says, "But the cubic has a quadratic term in it and its graphs are never parabolas." The first student suggests that they construct some tables. The partner agrees, and they make the tables shown in figure 2.9. At this time, both students point out that the values of $e(x)$ do become negative but then start increasing again for values *greater than 6*—"the graph of the last function isn't all there; it starts to go up again at the right." The students then conclude that the graph of any quartic function is "bowl shaped, something like a parabola, but flatter and wavier on the bottom."

These students would probably score 1 or 2 points on each of the "communicating mathematically" categories.

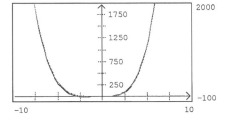

Fig. 2.6. Graphs of the first three examples of quartic functions

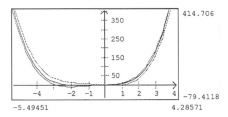

Fig. 2.7. Zoom of graphs of the first three examples

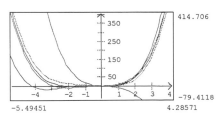

Fig. 2.8. Zoom of graphs of the first five examples

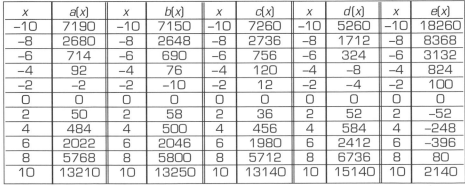

x	$a(x)$	x	$b(x)$	x	$c(x)$	x	$d(x)$	x	$e(x)$
–10	7190	–10	7150	–10	7260	–10	5260	–10	18260
–8	2680	–8	2648	–8	2736	–8	1712	–8	8368
–6	714	–6	690	–6	756	–6	324	–6	3132
–4	92	–4	76	–4	120	–4	–8	–4	824
–2	–2	–2	–10	–2	12	–2	–4	–2	100
0	0	0	0	0	0	0	0	0	0
2	50	2	58	2	36	2	52	2	–52
4	484	4	500	4	456	4	584	4	–248
6	2022	6	2046	6	1980	6	2412	6	–396
8	5768	8	5800	8	5712	8	6736	8	80
10	13210	10	13250	10	13140	10	15140	10	2140

Fig. 2.9. Tables of values produced for the five examples of functions

How would these students' work with the function-family exploration task be scored in terms of generating evidence and reasoning from this evidence? One possible rubric appears in figure 2.10. In this particular rubric, note that "different coefficient signs are chosen" is weighted more heavily than other categories. This decision reflects the teacher's particular emphasis; other teachers might allocate a proportionally different number of points to this category. According to this scoring system, the pair would receive points for their work as follows:

Quartic functions chosen:	2 points
Different coefficient absolute values chosen:	1 point
Different coefficient signs chosen:	3 points
Alternative representations used:	2 points
Representations manipulated:	2 points
Conjectures consistent with examples:	3 points
New examples based on old ones:	3 points

These scores reflect the central role that the alternative representations had in their reasoning process as well the fact that the two students neither considered all possible combinations of coefficient size and magnitude nor explained why they saw no need to do so.

Areas of Interest	Points	Qualities of Work
Generating evidence		
Quartic functions are chosen.	2	All examples are quartic.
	1	Some are quartic.
	0	None are quartic.
Different coefficient values are chosen.		Coefficients might include three types of numbers: (1) noninteger values, (2) very large and very small numbers, (3) numbers close to zero:
	3	All three types of numbers
	2	Any two of the three types
	1	Any one of the three types
	0	Only one or zero examples
Different coefficient signs are chosen.		Fourteen possible combinations of coefficients include + or – for each of five terms and 0 for each of four terms:
	5	+ or – for all terms, 0 for some terms
	4	+, –, or 0 for at least four terms
	3	+, –, or 0 for at least three terms
	2	+, –, or 0 for at least two terms
	1	+, –, or 0 for at least one term
	0	All + or all – for all terms

Areas of Interest	Points	Qualities of Work
Generating evidence		
Alternative representations are used.	2	Alternative representations are used as a significant part of the reasoning.
	1	Alternative representations are created.
	0	No alternative representations are present.
Representations are manipulated.	2	Manipulated representations are a significant part of the reasoning.
	1	Representations are manipulated.
	0	No representations are manipulated.
Reasoning from evidence		
Conjectures are consistent with examples.	3	Always consistent
	2	Usually consistent
	1	Somewhat consistent
	0	Never consistent
New examples are based on old ones.	3	Always related
	2	Sometimes related
	1	Rarely related
	0	Never related

Fig. 2.10. Potential scoring rubric for function-family exploration task

Probing Understanding. Assessment activities like the foregoing are useful in determining what students take away from their classroom experiences in algebra. They are also helpful in furnishing insight into students' understanding of mathematical exploration as a process in ways that are helpful to teachers in fine-tuning their instruction. There are also times when teachers want to know more than whether a student can do something, recall something, or apply something—there are times when teachers want to push the student's understanding to its limits.

Clinical and task-based interviews are particularly valuable tools for probing the depth and breadth of a student's understanding. These interviews involve giving students relevant but challenging mathematical tasks. These tasks should be likely to elicit from the student the mathematical ideas that the teacher wants to assess. The tasks should also be sufficiently challenging; students should need to think about the tasks and draw on their mathematical understandings in novel ways. The interviewer-teacher should also have a series of follow-up questions prepared. These questions should build on the variety of possible things that a student might say or do. The questions should be arranged in series. Each question in a series helps to further identify how the student understands algebraic ideas.

Most important, the interviewer-teacher needs to have a very clear goal in mind. When the student does something unexpected, the interviewer-teacher should have a basis on which to make an immediate decision about how to follow up on the student's thinking. An interview plan, therefore, includes a small number of very clear objectives and a carefully chosen set of tasks with follow-up questions that are more than a collection of disjoint exercises presented orally to the student. The sample interview described in the following paragraphs illustrates these characteristics.

One reason to include the study of families of functions within school algebra is to allow students to use this information in building and defending mathematical models of real-world phenomena. To assess how well students can generate and evaluate mathematical models, teachers can use an interview that begins with the following audiotape situation.

Audiotape Situation

You have just been hired by New Sound Tapes and Records to handle customers' questions. The first customer you meet asks you to determine a way to find the amount of playing time remaining on a tape if the amount of tape still wound on the reel is known.

Constructing a model of this situation is the "task" for the interview. The goal of the interview is for the teacher to acquire the information needed to describe how the student creates and defends a mathematical model, including the student's use of computing tools, mathematical representations, and data-collection methodology. The student is first asked to address the customer's question. A general plan for the interview, including a list of potential follow-up, probing questions, appears below. Indented text indicates what the interviewer might say to the student. Text in brackets indicates places where the interviewer must choose what to say or to what to refer on the basis of the student's responses (verbal or graphic). A possible sketch of the situation, possible data, and a related scatterplot are shown in figures 2.11, 2.12, and 2.13, respectively.

Give a copy of the audiotape situation to the student, saying,

> *We are going to start with what is written on this sheet of paper. After you are finished reading it, please tell me, in your own words, what it means to you.*

Wait for the student to read the situation. The student may choose to read silently or aloud. Remind the student to describe in his or her own words what the reading means. If the student does not describe the customer's question, ask,

> *What is the customer asking someone to do?*

After the student identifies the customer's question, ask,

> *How would you go about answering the customer's question?*

Encourage the student to construct an answer to the customer's question. As the student begins answering, record what the student writes and draws and follow up in the following areas:

♦ *Student's written work.* If the student draws something but does not explain the drawing, ask,

> *What did you draw there?*
> *Could you explain how that helps you to answer the customer's question?*

◆ *Use of tools and representations.* If the student uses a computing tool (e.g., calculator, curve fitter) or a representation (e.g., picture, graph, table), ask,

　What are you entering?
　What are you doing?
　Explain to me how you knew to [do/draw/use] that [tool, graph, table, etc.].

◆ *Data used in generating the model.* If the student suggests collecting data, investigate how he or she would do that with questions like these:

　What would you measure?
　How would you measure it?

If the student starts using numbers as data, determine how the student arrived at those numbers:

　How did you know to use those numbers?
　Where would I get a number like that to use if you did not tell me what numbers to use?

◆ *Knowledge of the modeling process.* Be sure to ascertain how the student generated each model considered. After the student generates the model(s), ask,

　Please tell me how you came up with that [equation, function rule, graph, etc.].

◆ *Evaluation of model(s).* Ask,

　How do you know that your [equation, function rule, graph, etc.] describes the relationship between the amount of tape still wound on the reel and the remaining playing time?

If the student does not understand the task in terms of the organization of the tape and reels, show the student a sketch of the situation (fig. 2.11). Also show this figure to the student if she or he has no idea about how to proceed with the modeling task, asking,

　How does this picture relate to the customer's question?

If the student suggests collecting data then says he or she cannot go any further without the data, or if the student still has no idea how to proceed after seeing the diagram, show the data chart in figure 2.12.

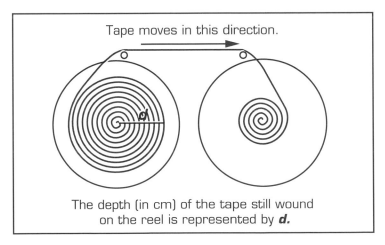

Tape moves in this direction.

The depth (in cm) of the tape still wound on the reel is represented by **d.**

Fig. 2.11. Diagram of tape movement on the tape reels

Depth of Tape Still Wound on Reel (in cm)	Playing Time Remaining (in sec)
2	1 350
4	3 400
6	6 150
8	9 600
10	13 750

Fig. 2.12. Potential data for the audiotape situation

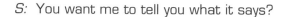

Then ask,

> *What might the information here have to do with the customer's question?*

Follow up later with these questions:

> *How do you think someone might have collected this information?*
> *Does this information make sense to you? Why?*

If the student still suggests no particular model or if the student does not use a graphic representation in his or her work, show the student the scatterplot (fig. 2.13). Then ask,

> *Does this make sense to you as a way to describe the movement of the tape?*
> *What does this tell you about possible answers to the customer's question?*
> *How might someone have obtained this information?*

The task and questions are all designed with a very clear goal: determining how the student constructs and evaluates mathematical models in a technology-intensive algebra environment. The task includes identifying what data to collect, how to collect and record these data, how to generate a model, and how to defend this model. A student might respond to this task in a one-on-one interview with the teacher-interviewer. The exchange between the student (*S*) and teacher (*T*) begins with the student silently reading the audiotape situation.

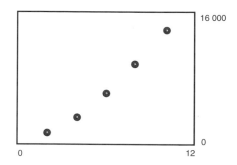

Fig. 2.13. Scatterplot of the audiotape-situation data

S: You want me to tell you what it says?

T: In your own words.

S: I'm working at a place where they sell tapes and my job is to answer people's questions. I guess I work in the complaint department. This customer comes in and asks me to find out how much playing time is left on the tape based on how, what would I say, like, thick the tape left on the first reel or spool is.

T: How would you go about answering the customer's question?

S: How would I find out how much playing time is left if I know how thick the tape on the spool is?

T: Is that what the customer is asking you to do?

S: I think so.

T: How would you find out how much playing time is left if you knew how thick the tape on the spool is?

S: Um, I'd have to do some measuring.

T: What would you measure?

S: I'd measure how thick the tape is and how much time is left.

T: I'm not sure that I understand how you would do that.

S: Well. The colored part here would be the tape and the white center is the spool. [Draws sketch in fig. 2.14a.] I'd measure this, from the spool to the edge. [Draws thick line segment in fig. 2.14b.] I'd mea-

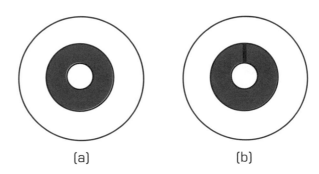

(a)   (b)

Fig. 2.14. Student's sketch of the situation

Assessment Matters: Time may not permit a teacher to interview students on a regular basis. However, this practice is clearly a very valuable window into how students think. Interviewing one or two different students each week would be a reasonable goal. It is common to find that the kinds of difficulties encountered by one student are shared by many classmates.

sure that in like inches or centimeters or something like that. Then I guess I'd have to play the rest of the tape and time how long it took for the rest of the tape to play out. However many minutes it took would be the remaining playing time. I'd probably have to use like a stop watch or something to time it. Then I'd have the time and the thickness.

T: I'm following you now. How would knowing the time and the thickness help you?

S: It wouldn't really, if I did it for just one tape. I'd probably do it for a couple of tapes. Different tapes may run differently.

T: May run differently?

S: Yeah. Different tapes would be different. Like one reel may have thinner tape because the tape had been played a lot and would be stretched out. So it would have more remaining playing time than another tape that was equally thick on the spool but newer. If I used different reels and got the measures from them then I could look for a rule.

T: A rule?

S: Yeah. Say I got all the numbers like this. [Draws chart in fig. 2.15.] Then I could say that the thicker the tape, the more playing time.

T: The thicker the tape, the more playing time?

S: Well, it should be. More tape left on the reel should mean more playing time left. That's what I was thinking when I made up these numbers. I thought since I couldn't actually measure the tape here I could at least write down numbers that would sort of fit what should happen.

T: Why did you choose those particular numbers?

Thickness	Time Left
1 inch	30 minutes
0.5 inches	15 minutes
0.25 inches	10 minutes

Fig. 2.15. Student's chart of hypothetical data

After the interview ends, the teacher would not attempt to "score" the student's response. Instead, the teacher would write a description of what the student did. An example of such a description appears in figure 2.16. This description would then become part of the student's portfolio, along with the student's written work during the interview and perhaps an audiotape or videotape of the student working on the modeling task. The interview supplies baseline data about the student's work early in the student's high school mathematics experience and can be compared with future modeling performance.

The student (S) described collecting data by measuring the "thickness" of the tape and using a timing device to determine how much time was left. S made up a data set that included an increasing relationship between the amount of tape left and the amount of playing time left. S said that her fictitious data values were intentionally not linear; as S said, "the numbers go down faster when the thickness is less because less tape is needed to make one ring around the spool on a thinner tape." S used a curve fitter and her fictitious data to generate quadratic, cubic, quartic, and exponential models of the relationship. S said that a linear function would not be appropriate because the "playing time doesn't go up as fast when there is very little tape left on the spool." In each case, the input and output variables were thickness and time remaining, respectively. S chose the quartic function as the best model because its goodness-of-fit value was best. When asked if any others were any good, S said that they all "looked pretty good" because "they go almost through the data points." S maintained that the quartic was the best because "the calculator told me so."

Note: S's blatant dependence on the goodness-of-fit value occurs in the absence of talk about the characteristics of the function. Several of S's classmates also depended on the goodness-of-fit value. We will address this in class during our next modeling activity.

Fig. 2.16. Teacher's summary of student's modeling work during the interview

CONCLUSION

The incorporation of technology into the algebra experience of high school students raises important questions about what students and teachers should do in and out of the classroom. A need exists for the roles of students and teachers to change and to blend as all use technology in the pursuit of mathematical understanding. The changes in the learning environment also suggest a need to reconsider how students' understanding of mathematics should be assessed and evaluated. The greatest challenges for instruction and assessment seem to revolve around a central idea: technology opens many doors and thereby drastically increases the number of things that teachers and students can do with respect to mathematics teaching and learning. Teachers therefore need to assess different mathematical understandings and carefully document what students do and know as they progress through not only the algebra strand but also the other strands of school mathematics. Furthermore, teachers need to reevaluate continually their pedagogical and assessment practices. It is likely that future technology will continue to have an impact on the school mathematics curricula, challenging and changing the perspective of what and how students learn.

CHAPTER 3
A FUNCTIONS APPROACH TO ALGEBRA

Functions are a central and unifying concept of school mathematics. With the help of technology, the notion of function can be expanded from rules to compute output values from given input values to a dynamic study of the relationships between two or more quantities that vary. With this expansion of the concept, a functions approach to algebra becomes not only possible but actually appropriate. Functions can be readily used and studied by high school students in the context of mathematical modeling, in identifying and characterizing families of functions, in understanding systems of functions and relations, in exploring dynamical systems, and in meaningful discussion of symbolic manipulation. As the following discussion of these topics illustrates, a functions approach to algebra can be powerful. This approach not only helps students connect algebra to the real world and probe mathematical structures but also serves as a link among the multitude of algebra topics that students encounter in a technological world.

MATHEMATICAL MODELING

Mathematical modeling is a natural tool for understanding relationships among quantities. It offers a way for a highway engineer to determine the amount of asphalt needed to pave a roadway of a given length or for an economist to theorize about the effects of a policy of the Federal Reserve Board on the economy. Students who understand functions as relationships between quantities have a powerful way of explaining real-world phenomena.

The modeling process involves several central steps: determining the need for a mathematical model, identifying important factors (variables) to include in the model, determining how to obtain relevant data, constructing an accurate model, choosing among competing models, and reasoning about the situation from the chosen model(s). In a functions approach to algebra, the need for a model that relates data representing quantifiable variables suggests the potential use of a function to model the situation. Students draw on their knowledge of functions—as well as on knowledge of the real-world situation and sometimes on knowledge of other mathematical ideas—to generate a function model. This process can have one of several forms. The next three examples illustrate three different ways in which students might generate, choose, and use function models. Three classroom-ready activities that correspond to the three examples can be found at the end of this chapter.

Fig. 3.1. Illustration of molding around a rectangular window

Example 1
Modeling by Applying Known Rules: Window Moldings

A common name for the narrow rim of wood that surrounds the edge of a window, door, or floor is *molding*. The shaded region in figure 3.1 illustrates the molding around a rectangular window. A rough first approximation of the amount of molding needed for a window is the perimeter of the window. A model of the approximate length of molding in feet as a function of the length (L) in feet and the width (W) in feet of a rectangular window would be given by the rule

$$P(L, W) = 2L + 2W.$$

To arrive at this function model, students need only adapt the formula they already know for finding the perimeter of a rectangle on the basis of the length and width. So, one way in which algebra students may generate function models is through the use of accepted formulas. This method of finding function rules to use as models can be extended to include the generation of models to approximate the amount of molding needed for other windows, such as the hexagonal window shown in figure 3.2.

After obtaining the rule for function *P,* students should discuss the quality of their model. In this example, they may talk about their confidence or lack of confidence in the model. Function *P* does yield the perimeter of the window opening. However, the value for the perimeter obtained from the rule is subject to input errors from measuring the length and width of the window. A more important limitation of this particular model is that it yields the inside perimeter of the molding rather than the outside perimeter, which is needed to know how much molding to buy. It is necessary to know the outside perimeter because installing the molding requires removing corner pieces from the ends of each side piece, as shown in figure 3.3.

Fig. 3.2. Illustration of molding around a hexagonal window

Fig. 3.3. Installing molding by removing the corner pieces

Students who recognize the strengths and the limitations of the model can then intelligently use function *P* as a basis to analyze and possibly answer questions about window molding:

♦ How much molding would be needed for a rectangular window that measures 4 feet by 2.75 feet?

♦ What dimensions could a rectangular window have if it takes approximately 28 feet of molding to surround the window?

♦ If the width of a rectangular window remains constant, how will changing the length of the window affect the amount of molding needed to surround it?

♦ How does the function *P* differ from a function that models the approximate amount of molding needed to surround a hexagonal window? Explain why these differences make sense.

Using previously encountered formulas as a basis of a mathematical model may be the fastest way to obtain an acceptable model, but it is not necessarily the most interesting or the most powerful use of mathematical modeling. However, this molding situation illustrates the essence of mathematical modeling—reasoning in ways that make sense mathematically without losing sight of what is realistically sensible, and vice versa.

Assessment Matters: The questions listed can be used for assessment as can the "Try This" activities that follow. Another good assessment question is the following: You need to purchase four-inch-wide molding for a rectangular window so that the inside dimensions are 3 feet by 4 feet. How many feet of molding should you purchase?

Try This: In the window-molding situation, although an approximate answer can be calculated from the given information, additional information is needed for more accurate answers. Ask students to identify the missing information, generate reasonable values for the information, and answer the questions. Have them create similar situations, function rules, and questions.

Try This: Write a rule for the length M of m-inch-wide molding needed to trim a rectangular window that is x feet by y feet.

Example 2
Mathematical Modeling through Deductive Reasoning: The Pet Wards

A second way in which students can construct function models is through deductive counting processes. This method of generating function models is especially helpful in exploring the pet-ward construction problem, adapted from *Concepts in Algebra: A Technological Approach* (Fey and Heid 1995).

Pet-Ward Construction Situation

A national pet-hotel chain is planning to build units for a series of franchises. Each unit for small pets is a row of two-meter-by-two-meter square wards. The wards are connected as shown in the partial floor plan below:

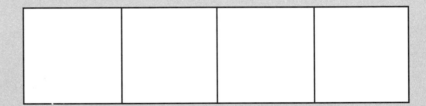

Walls for these units come only in two-meter panels, and the number of two-meter panels needed depends on the number of wards to be included in the unit. Because the management plans to build many units of different sizes, the manager wants to have a rule relating the number of wards and the number of panels.

Students might generate the desired function rule for the number of wall panels as a function of the number of wards in several ways. Figure 3.4 illustrates one way students might envision the arrangement of wall panels. From this view, students might count n wall panels at the top of the row of wards, n wall panels at the bottom row of wards, and $(n + 1)$ vertical wall panels that form the walls between and at the end of the wards to arrive at a total of $n + n + (n + 1)$ wall panels. Their function model P would be given by the rule

$$P(n) = n + n + (n + 1).$$

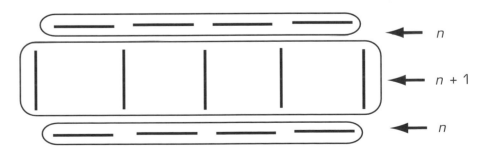

Try This: Pet wards are arranged in k rows with six wards in each row. How does the number of panels needed vary as a function of k?

Fig. 3.4. Pet-ward wall panels grouped as top, bottom, and side walls. The total number of wall panels is $n + n + (n + 1)$.

Another option is to think of the pet-ward wall panels grouped as in figure 3.5. The number of wall panels needed for *n* wards would be three wall panels for each *n* ward and one extra panel to close the left end of the row of wards. The total number of panels here would be

$$1 + (3n).$$

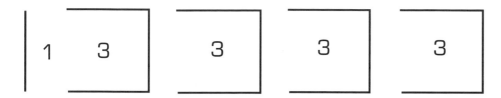

Fig. 3.5. Pet wards with 1 + 3n wall panels

Students could use several other counting strategies to arrive at equivalent rules.

Regardless of how they deduce the model, students obtain a function that relates the number of wall panels to the number of wards. This model then can be used for reasoning about the pet-ward situation. Questions and tasks students may pursue here include the following:

♦ Use your rule to determine the number of wall panels to be shipped to a location where twenty wards are to be built.

♦ Is it true that a row of wards twice as long as a given row will require twice as many wall panels as required in the given row?

♦ How does the number of wall panels change as the number of wards built increases?

Answering such questions as the last one connects naturally with students' understandings of linear functions, which will be discussed later under "Families of Functions."

The last two situations—the window molding and the pet wards—and the methods for constructing function models exemplified by them were similar in two ways. First, they led rather directly to a single model. Second, algebra students are likely to have acquired previously an adequate understanding of the mathematics needed to develop the models. As a result, they would generate models on which anyone would agree mathematically as long as they accepted the same assumptions about which quantities to consider. In the molding situation, a formula yields perimeter on the basis of length and width. Algebra students would know this formula from their previous mathematical experiences. As long as students agree that the perimeter of the window is an adequate estimate of the amount of molding needed, they should all agree that function P with rule $P(L, W) = 2L + 2W$ is an acceptable model. In the pet-ward situation, students' construction of rules that differ in appearance, such as $P(n) = n + n + (n + 1)$ and $p(n) = 1 + 3n$, can lead to a lively discussion of what it means for function rules to be equivalent. Students who use valid counting techniques will always generate a rule equivalent to $p(n) = 1 + 3n$. The next example considers what can happen when the modeling activity is more open-ended than these last two examples.

Teaching Matters: Teaching mathematics is a rich and exciting experience when students and teachers see connections both to the past and to the future. The set of questions proposed for the pet-ward construction situation includes a look backward to the meaning of proportion and acts as a precursor to the concept of rate of change.

Example 3
Mathematical Modeling through an Analysis of Real-World Situations and Curve Fitting: The Fuel Bills

In Iowa City, Iowa, residents receive combined gas and electric bills. Each monthly bill gives the customer the daily cost and the daily consumption of gas and of electricity and the average temperature during the month. Information taken from a new homeowner's monthly bills from August 1992 through March 1993 appears in figure 3.6. On seeing the bill, the customer asked the following questions: Does a relationship exist between the daily cost of fuel and the average temperature for a given month? If such a relationship exists and a function can model it, the customer could estimate her daily cost of gas and, therefore, her monthly gas costs. This information would be valuable to her as she tries to establish a household budget. Developing a function model would be worth her effort.

Month Ending	Average Daily Cost of Gas	Average Daily Cost of Electricity	Average Monthly Temperature (°F)
21 March 1993	$1.76	$0.73	27
22 February 1993	$1.97	$0.80	27
19 January 1993	$1.94	$0.65	23
21 December 1992	$1.86	$0.86	31
18 November 1992	$1.41	$0.77	44
17 October 1992	$0.78	$0.82	58
16 September 1992	$0.38	$0.78	69
17 August 1992	$0.35	$0.98	70

Month Ending	Average Daily Use of Gas (CCF)	Average Daily Use of Electricity (KWH)	Average monthly temperature (°F)
21 March 1993	3.6	6.7	27
22 February 1993	4.0	7.5	27
19 January 1993	3.9	5.7	23
21 December 1992	3.6	7.8	31
18 November 1992	2.5	7.0	44
17 October 1992	1.2	7.7	58
16 September 1992	0.4	6.5	69
17 August 1992	0.4	8.5	70

Fig. 3.6. Information from monthly fuel bills

Because her home had gas heat, the customer believed that the daily cost of gas should depend on the average temperature. Furthermore, she felt that more gas would be used on colder days because more heat would be needed to warm the house. She concluded that the daily cost of gas should increase as the average temperature decreased. She created a scatterplot as shown in figure 3.7 and used a computing tool to generate the fitted linear function for cost as a function of temperature to obtain the rule

$$c(t) = -0.035\,356\,0t + 2.848\,65.$$

The customer then graphed the fitted function (see fig. 3.8).

The customer wanted to get an idea of what her April gas bill might be. Using computing tools to trace the fitted function graph and using 55°F as the predicted average temperature for the month, she located two points with coordinates (54.890 1, 0.907 96) and (55.439 6, 0.888 534). Since 55 was between 54.890 1 and 55.439 6 and the function was linear, she knew that $c(55)$, or the average daily cost of gas for an average temperature of 55°F, would be between 0.907 96 and 0.888 534. The customer concluded that her daily cost would be about $0.90 and that her April gas bill would be about $27.00, or 0.90×30.

The customer decided to repeat her reasoning to find her cost of electricity. The scatterplot and the fitted function graph that she obtained appear in figure 3.9; the rule for this fitted function is

$$e(t) = 0.002\,986\,81t + 0.668\,450\,002\,3.$$

The homeowner immediately rejected this model as unreasonable! Why? First, this model suggests that the cost of electricity would increase as temperature increases. Moreover, the points of the scatterplot do not appear to lie in a linear pattern, and they do not seem to lie very close to the graph of her fitted function. The mathematical merit of this model is very questionable. In fact, when she generated fitted quadratic, cubic, quartic, and exponential functions (fig. 3.10), she found that these potential models seemed to be as weak as the linear model.

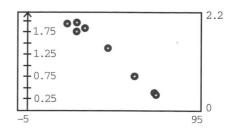

Fig. 3.7. Scatterplot of the daily cost of gas as a function of average monthly temperature (in °F)

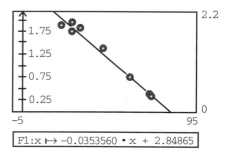

Fig. 3.8. Graph of the fitted linear function for the daily cost of gas as a function of average monthly temperature (in °F)

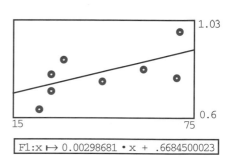

Fig. 3.9. Graph of the data and the fitted function for the daily cost (in $) of electricity as a function of average monthly temperature (in °F)

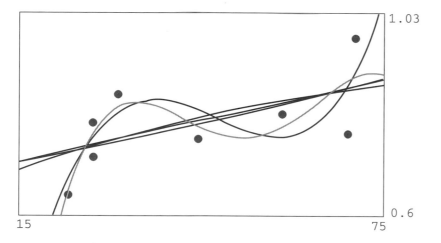

Fig. 3.10. Graphs of all fitted functions (linear, quadratic, cubic, quartic, and exponential) for daily cost (in $) of electricity as a function of average monthly temperature (in °F)

When she thought more about her home, the customer admitted that even trying to generate a function model was an unrealistic task. She uses some electricity to air condition the house but virtually none for heating the house. Her typical uses of electricity involve lights, computers, television, and radios. Although it might be true that she would use

Assessment Matters: The fuel bill data is for August to March. Probe students' thinking by having them describe how the functions relating gas and electricity consumption and cost to average temperature would be affected if all twelve months were included. What if April to July were considered separately?

Teaching Matters: In the fuel-bill modeling problem, students are given data and asked to produce a model. Another important phase of the modeling process is for students to generate a model given the situation. Data-gathering tools like the Texas Instruments CBL System can enhance students' facility with identifying variables they think are related and gathering data to refine and test their conjectures. Students can use these tools to measure motion, temperature, sound, light, pH level, and more as they change over time. Through simple experiments, students can gather real-world data for linear, exponential, logistic, and periodic relationships, to name a few.

some of these appliances more in the winter when daylight hours are fewer, there is no reason to believe that the daily temperature would be very strongly related to her consumption of electricity. Furthermore, she also knows that the electric rates changed substantially between August 1992 and March 1993. It just does not make sense for this homeowner to think about her daily cost of electricity as a function of average temperature.

Her realization of the improbability of a relationship between daily cost of electricity and average temperature led her to reconsider her model of daily cost of gas and average temperature. Would she get a better result if she considered daily consumption of gas as a function of average temperature? After all, daily consumption would not be subject to rate changes in the same way as daily cost would be. It seemed worth a try. The new function model that she constructed for daily consumption as a function of daily temperature has the rule

$$G(t) = -0.079\ 493\ 2t + 5.917\ 89.$$

The scatterplot and graph of this fitted function appear in figure 3.11.

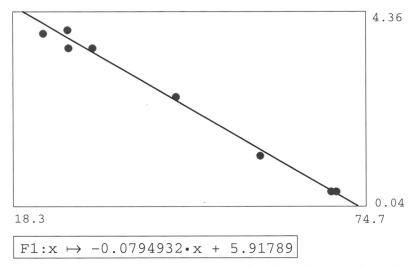

$$F1: x \mapsto -0.0794932 \cdot x + 5.91789$$

Fig. 3.11. Scatterplot and graph of the fitted linear function for average daily consumption of gas (in CCF) as a function of average monthly temperature (in °F)

Judging from the position of the data points compared with the graph of the fitted function, this model seems to be mathematically reasonable. The customer could also check the goodness-of-fit value for this model (0.141 47) and the goodness-of-fit value for the average-daily-cost-of-gas model (0.086 462 9). These values suggest that the model for consumption is a more mathematically sound model than the model for cost. This observation is also supported by the graphs of these fitted functions with their respective scatterplots. The data points lie close to the graph for the consumption-based function. However, both values are clearly closer to 0 than to 1. This numerical observation, as well as the overall reasonableness of both graphs as summaries of the scatterplots, indicates that both models are fairly good from the purely computational point of view.

In the real situation, the model suggests that as temperature increases, the daily amount of gas used decreases. That, too, makes sense. The line does not fit the points perfectly, which is not surprising. The customer also has a gas clothes dryer. There is obviously some "noise" in the data because some of the gas is consumed to dry clothes and not to heat the home. It may also be true that because more clothing is worn in colder weather, more clothing is dried in the dryer; therefore, more gas would be used when the temperature is lower. Also, it may take somewhat more gas to heat the dryer to the necessary temperature on colder days than it does on warmer days. All this information suggests that both functions G and c make sense mathematically and situationally as models that the homeowner could use to better understand how her use of gas and outside temperatures could be related.

This discussion of modeling based on the information shown on a customer's fuel bill is easily within the reach of students when algebra is taught and learned from a functions approach. The homeowner needed to understand function as a relationship between quantities that vary. She also used her knowledge of linear functions, and other families of functions, to interpret the meaning of trends and particular coefficients involved in the potential models she considered. She used a computing tool to generate the fitted functions. She then evaluated the quality of her potential models on the basis of two notions about fitted functions: the visual relationship between a function that fits well and the data in a scatterplot and the interpretation of goodness-of-fit values. All these ideas are now within the reach of algebra students.

FAMILIES OF FUNCTIONS

Many families of functions can be studied and used by high school algebra students. A knowledge of important families, as suggested in the preceding mathematical modeling section, is useful in constructing and evaluating models. Understanding function families also builds a basis for the study of mathematical structures, an opportunity to develop mathematical reasoning, and a foundation for further mathematical explorations and applications.

Several families of functions are useful both in modeling mathematical activities and in generating fruitful mathematical explorations. Students should engage in several different types of function-family activities. The following discussion presents four particularly useful and interesting families—linear, exponential, rational, and polynomial. The student activity sheets at the end of this chapter illustrate different types of classroom activities through which students might explore and apply their knowledge of families of functions. These function families are further developed in *Concepts in Algebra: A Technological Approach* (Fey and Heid 1995).

Fundamental Function Families

Linear functions (for example, f(x) = ax + b)
Linear functions are powerful tools for understanding real-world phenomena. Many problems from such diverse fields as biology, engineering, and economics can be solved through the application and analysis of linear functions. Linear functions are unique in that they are the only functions with constant rates of change. That is, equally spaced input values yield equally spaced output values, both graphically and numerically.

Linearity is a familiar concept for learners of all ages. Most students begin early in school, or even before school, to build an experience base for linearity through some fairly simple problem solving. For instance, the task of counting the number of shoes in the classroom can be accomplished by counting the number of pairs, that is, by counting by twos. This and other skip-counting activities are a natural building block toward

Try This: A good starting exercise with the Texas Instruments CBL System is for students to graph their motion as they move away from and toward a motion detector. For example, the motion of step forward, stop, step forward, stop, step back, stop, step back might be seen as the following graph:

Have students form teams. One member from each team generates a motion to be graphed. Try having students generate their motions while other students close their eyes. The graph of the motion can be shown to the students who did not see it, and they can try to duplicate the graph though motions of their own.

understanding linearity and the constant rate of change characteristic of linear functions. In the shoe-counting problem, for example, the constant rate of change, 2, can be thought of as two shoes in each pair; the natural starting point (i.e., the shoe-intercept) is zero shoes in zero pairs. Reconsidering early experiences like these leads nicely into recognizing linear functions in the real world as well as analyzing the characteristics of linear functions through their graphs, tables, rules, and other representations.

The following loan-repayment problem suggests ways to facilitate linking students' knowledge of arithmetic with early algebraic notions of linearity. Early subtraction and division approaches to the problem can be revisited later through tabular, graphical, and symbolic representations.

> Loan-Repayment Problem
>
> Determine the number of months required to pay off a $200, no-interest loan from your mother. She expects to receive a monthly payment of $15.

Representations of portions of different problem solutions appear in figure 3.12.

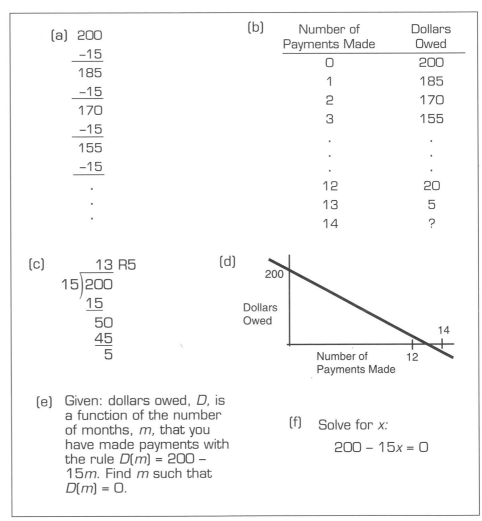

Fig. 3.12. Possible solution methods for the loan-repayment problem

Students should recognize the effects on each of these representations of changes in the values of *m* or *b* when the functions are presented as rules of the form $L(x) = mx + b$. They should also be able to compare and to modify two or more linear-function models of real-world phenomena.

Early work with functions often requires students to find only a single output value or a single input value for a function rule. To help students access the full power of functions, it is necessary to shift their attention away from a point-by-point analysis and toward a recognition of overall patterns and trends in linearly related data. The discussion of the following job-offer situation, adapted from *Concepts in Algebra: A Technological Approach* (Fey and Heid 1995), illustrates one way to begin this shift in focus.

Job-Offer Problem

Given the two job offers below, determine the *better-paying* summer job. Explain your reasoning.

Offer 1: At Timmy's Tacos you will earn $4.50 an hour. However, you will be required to purchase a uniform for $45.00. You will be expected to work 20 hours each week.

Offer 2: At Kelly's Car Wash you will earn $3.50 an hour. No special attire is required. You must agree to work 20 hours each week.

If we consider wages earned as a function of the number of hours, or the number of weeks, worked for each job offer, algebraic-function rules could be written to represent those relations. These rules can be used to produce computer-generated function tables and graphs for wages earned over time. Analyzing these algebraic-function rules, tables, and graphs may contribute to an understanding of the rate at which wages increase for each job offer as the number of hours worked increases. Numerical and graphical "models" of the behavior of the two functions can serve as tools to aid in selecting the better-paying summer job.

Consider wages earned as a function *D* of the number of hours *h* worked. For the first offer, we could express this relation as $D_1(h) = 4.5h - 45$. For the second job offer, write $D_2(h) = 3.5h$.

Function tables for these two rules appear in figure 3.13.

Number of Hours Worked	Number of Weeks Worked	Wages Earned Job 1	Wages Earned Job 2
0	0	−45	0
20	1	45	70
40	2	135	140
60	3	225	210
80	4	315	280
100	5	405	350
120	6	495	420
140	7	585	490
160	8	675	560

Fig. 3.13. Table of values for the job-offer problem

Try This: When working with tabular representations of functions, try adjoining columns of related data and asking students to analyze relationships among several sets of variables. A possible extension activity follows.

In the job-offer situation, we have expanded the table to include the column "Number of Weeks Worked." Would we expect to make any changes in the rules to express wages earned as a function of the number of weeks worked? If yes, explain your answer and state the new rules.

Would we obtain the same wage-function graphs if we considered the number of weeks, instead of the number of hours, worked? Experiment by using computers or calculators to generate function tables and graphs. Explain your reasoning.

Students can roughly sketch the overall shape of each wage-function graph suggested by the data. From the sketches they can note overall trends. For example, the output values for function D_1 increase at a rate of 4.5 for each additional hour worked. The intersection of the graph of $D_1(h) = 4.5h - 45$ with the vertical axis is -45. It gives the amount of money owed at the start because of the uniform purchase. The intersection of the graph of $D_1(h) = 4.5h - 45$ with the horizontal axis is 10, which is the number of hours that would have to be worked to break even.

Computer- or calculator-generated function tables and graphs can help in the successive approximation of solutions to the following equation and inequality:

$$4.5h - 45 = 3.5h$$
$$4.5h - 45 > 3.5h$$

Superimposing the graphs of the functions D_1 and D_2 suggests that the jobs will pay the same for somewhere between forty and fifty hours. (See fig. 3.14a.) Zooming in on this graph (see fig. 3.14b) suggests that the jobs pay the same for forty-five hours and that job 1 pays more if more than forty-five hours are worked.

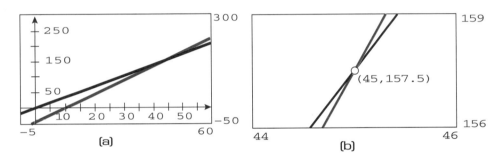

Fig. 3.14. Graphical representations of the job-offer problem

Activities like "Exploring Linearity" (Activity 6) at the end of this chapter serve several purposes in developing an understanding of linearity. First, they engage students in considering the advantages and disadvantages of different representations—graphical, numerical, and symbolic—for linear functions. Second, they give students some experience in considering differing rates of change. In developing an understanding of linearity, it is important that students have a range of activities that expose them to working with a variety of linear functions that vary in slope and intercepts. Through working with a variety of these functions in applied contexts, students can develop an intuition about linearity and exemplars with which they can compare other families of functions. As students work with situations that involve constant rates of change, they will grow in their ability to recognize and discuss linear functions as they arise in other areas of mathematics.

The job-offer problem gave enough direct information to allow students to generate a function rule. More often in real-world situations, students will need to gather the data themselves and then generate a relevant rule. Computer simulations, such as the one described in the next section, can help prepare students for full-blown mathematical modeling tasks.

The Geometer's Sketchpad (Jackiw 1991), one of a growing stock of geometric investigation software, allows the dynamic exploration of geometric relationships. An important feature of the software is that it enables the user to gather and tabulate data automatically from these experiments. This type of computer software significantly contributes to the integration of geometric reasoning with numerical, graphical, and symbolic reasoning. The following example illustrates how one teacher used the Geometer's Sketchpad to develop a linear function relating circumference to the radius of a circle.

In figure 3.15, the Geometer's Sketchpad has been adapted to create an elementary but dynamic model of the relation between the radius and the circumference of a circle. As students work with dynamic models like these, issues of measurement error and technology-generated round off will arise. (We will return to this model in activity 9, which engages students in working with the relation between the radius and the area of a circle.)

Teaching Matters: When computing tools are used to generate numerical data, round off is a common consideration that gives students an opportunity to think about the effect of computer round off. Have students try the radius and circumference activity and compare their results with those obtained using the rule C(r) = 2πr. Ask them to explain any discrepancies.

Radius(Circle 1) = 0.70 inches
Circumference(Circle 1) = 4.39 inches

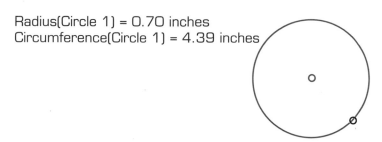

Radius(Circle 1)	Circumference(Circle 1)
0.91	5.75
0.20	1.25
0.30	1.89
0.61	3.85
0.70	4.39

Fig. 3.15. Using a geometric investigation tool to explore the relationship between the radius and the circumference of a circle

By moving the cursor or pointer back and forth along the horizontal bar, we can observe changes in the size of the circumference of the circle as the radius increases or decreases. A running counter also displays the numerically resulting changes in the radius and the circumference of the circle. Data collected from this dynamic model, paired with data collected from physical experimentation with circular objects, can be used to arrive at a meaningful linear model defining the circumference C of a circle as a function of the radius r. Students can be asked to consider the "goodness of fit" of their model for the relation between the radius of a circle and its circumference.

Activity 9, "Circumference and Area," at the end of this chapter engages students in a Geometer's Sketchpad activity that requires comparing a linear function with a function from a nonlinear family of functions. This problem not only connects geometry and algebra but also engages students in using a different (geometry based) form of technology in an

Assessment Matters: The Geometer's Sketchpad allows the teacher to access script files and examine the steps that the students use when they attack a problem. This information offers insights into their levels of understanding and their problem-solving ability.

algebra setting. Students have the opportunity to collect, display, and interpret data that they themselves generate. Reasoning about why the relationships are or are not linear forces them to integrate notions of geometry and algebra.

Exponential functions (like E(x) = CA^x)

Exponential functions form another important family. Exponential functions are extremely useful in such well-known applications as population growth, compound interest, radioactive decay, and value depreciation. Exponential functions are monotone like linear functions, but they do not have constant rates of change. In fact, the rate of change of an exponential function is either continuously decreasing or continuously increasing.

We can begin a consideration of exponential growth by pointing out that not all mathematical functions share the properties defining linear functions and by asking students to consider the relation between time and wages earned in the following situation.

The Doubling Pay Schedule

An eccentric but hypothetical employer pays you wages as determined by the pattern suggested in the following table. You earn one penny on day 1. Each day thereafter, you receive double the amount of the previous day's earnings.

Day d	Earned Wages on Day d
1	.01
2	.01 + .01 = .02
3	.02 + .02 = .04
4	.04 + .04 = .08
5	.08 + .08 = .16

What would you expect the wages to be on day 22?

Would you expect to use a linear function to represent the relation suggested by the data? That is, could you represent the wages as a function of the number of the day worked by a rule of the form $W(d) = md + b$? Explain your reasoning.

It is informative for students to compare an extended version of the table in the problem with the wage function defined by the rule $W1(d) = 2d$ (see fig. 3.16). They can compare rates of change in earned wages as time passes for these two pay schedules and begin to recognize essential differences between linear and exponential functions.

Students can also consider the following alternative tabular representation of the (time worked, wages earned) data.

Day d	Earned Wages on Day d
1	.01
2	.02 = .01 + .01 = 2 × .01
3	.04 = .02 + .02 = 2 × .02
4	.08 = .04 + .04 = 2 × .04
5	.16 = .08 + .08 = 2 × .08

Day d	$W(d) = ?$	$W1(d) = 2d$
1	.01	.02
2	.02	.04
3	.04	.06
4	.08	.08
5	.16	.10
6	.32	.12
7	.64	.14
·	·	·
·	·	·
·	·	·
21	10 485.76	.42
22	20 971.52	.44
23	41 943.04	.46
24	83 886.08	.48

Fig. 3.16. Earned wages on day d

Thus, wages earned on day d (after the first day) are determined by wages earned on the preceding day. To find the wages for any one day, we need only double the wages for the previous day. We can state this pattern using symbols:

$$W(2) = W(1) \times 2,$$

$$W(3) = W(2) \times 2,$$

$$W(4) = W(3) \times 2,$$

and so on. That is, $W(d + 1) = W(d) \times 2$.

It is often easier for students to work with exponential situations by using the relationship between successive function values; in fact, their first recourse in evaluating exponential functions is often through calculating the next value from the previous value. Rules whose evaluation depends on knowing the value preceding it are called *recursive*. The replay capabilities on graphing calculators permit recursive rules to be easily evaluated.

It does not take long, however, for students to realize that even with instant-replay capability, recursive rules do not efficiently determine output for very large input values. For example, they would quickly find out that they need a lot of values to find the wages for day 50 of the wage-doubling function if they know only the recursive definition. Once students understand this difficulty, they will be ready to appreciate that the function rules for recursive functions can be stated in an alternative and arguably more usable way.

We can think about this wage-doubling function as a function for which we know that the wages start at one penny each day and that the wages for any day are two times what they were for the preceding day. To get the wages for the second day, multiply the wages for the first day by 2. So $W(2) = 0.01 \times 2$. To get the wages for the third day, multiply the wages for the second day by 2. So $W(3) = (0.01 \times 2) \times 2 = 0.01 \times 2^2$. Using similar reasoning, we find that the wages for the fourth day are 0.01×2^3, and so on, leading to the conclusion that the wages for the

Try This: Pose the task of comparing the wage-doubling function with the function **W1(d) = 2d.** Suggest that the students make a sketch that illustrates differences in how the two functions grow. Have them discuss their representations in small groups.

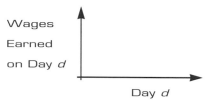

Wages
Earned
on Day d

Day d

Assessment Matters: As students work in small groups, a teacher has the opportunity to observe their ability to communicate mathematics to others and to cooperate with others to solve problems. The dispositions of students toward mathematics, such as their levels of enthusiam, perseverance, and motivation, can also be observed.

Teaching Matters: Have small groups of students discuss the strengths and weaknesses of the recursive form of a function as compared to the closed form. The ease of computation for large input values clearly is a point in favor of the closed form, but which form is more natural or more useful in other types of applications?

dth day are $0.01 \times 2^{d-1}$. This expression supplies an alternative rule for the wage-doubling function.

Functions whose first term is C and whose other terms are generated by multiplying the preceding term by a constant, a, can be expressed in the form $f(n) = Ca^{n-1}$, where n is the number of the term. Moreover, for any function of the form $f(n) = Ca^{n-1}$ with positive base a and for any integers n and $n + 1$ in the domain of the function, it is true that $f(n + 1) = a \times f(n)$. By contrast, recall that successive terms described by linear functions can be generated by adding a constant to the previous term. For a linear function g, the slope of its graph is closely related to how the output changes as the input changes from x to $x + 1$. That is, $g(x + 1) = g(x) + m$, where m is the slope of the linear function.

Exponential decay may be introduced through such modeling activities as the paper-folding task illustrated in figure 3.17. Take a large sheet of paper, such as a sheet of newsprint, and fold it in half horizontally. When the folded paper is placed on a flat surface, the size of the "top sheet" is half the size of the original sheet. Fold the paper in half horizontally again; the size of the "top sheet" of the folded paper is now one-quarter the size of the original sheet. Continuing with this folding process, students get a first look at how the size of the "top sheet" continues to become smaller. Also, the amount by which the size decreases at each step decreases rapidly at first (from 1 to 1/2 to 1/4) and then by much smaller amounts as the size itself becomes smaller, that is, from 1/4 to 1/8 to 1/16. The paper-folding experience furnishes a physical model of the constantly decreasing rate of decrease—a phenomenon that is more easily experienced than said or read!

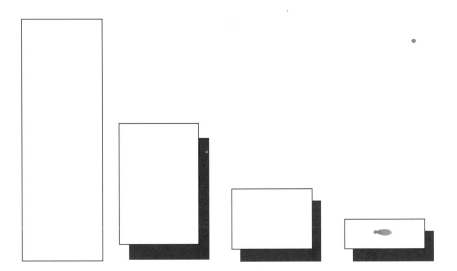

Fig. 3.17. An illustration of the rapidly and then more slowly decreasing size of the "top sheet" of the folded paper

The family of functions used to describe the patterns of change in the wage-doubling situation or the paper-folding situation are called *exponential functions*. Properties of exponential growth and decay can be studied by examining exponential functions of the form $f(x) = Ca^x$.

The following situation, adapted from *Concepts in Algebra: A Technological Approach* (Fey and Heid 1995), might be modeled by an exponential function, but the question being posed requires a somewhat different interpretation of the meaning of an exponent.

Swimming-Pool Bacteria

If the bacteria count in a heated swimming pool is 1500 per cubic centimeter on Monday morning at 8 A.M. and the count doubles each day thereafter, what is the bacteria count at 2 P.M. on Thursday, 3.25 days after the initial count?

The question being asked requires the use of a noninteger exponent. For integer inputs, we could fairly easily determine corresponding outputs through an application of the rule $N(d) = 1500 \times 2^d$ where d is the number of days since the initial count and $N(d)$ is the number of bacteria after d days. That is, three days later the bacteria count is given by the equation $N(3) = 1500 \times 2^3$.

The repeated multiplication process used to evaluate $N(3) = 1500 \times 2^3$ does not work when the exponent is not an integer. How can we determine $N(3.25)$? Does it seem reasonable that $2^{0.5}$ should be between $2^0 = 1$ and $2^1 = 2$? Should $2^{3.25}$ be between $2^3 = 8$ and $2^4 = 16$? In general, for an exponential function where the base a is positive, but different from 1, how is $f(x)$ related to $f(p)$ and $f(q)$ where $p < x < q$?

Fortunately, we can determine output values for all real input values, including nonintegers, by using computer- or calculator-generated tables and graphs. Consider the computer-generated table and graph for the bacterial-growth function in figure 3.18.

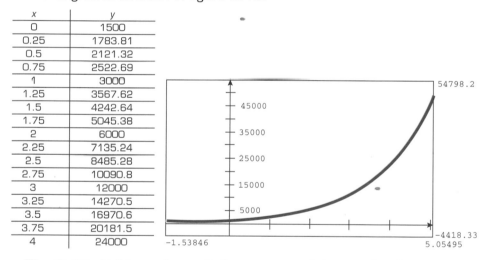

x	y
0	1500
0.25	1783.81
0.5	2121.32
0.75	2522.69
1	3000
1.25	3567.62
1.5	4242.64
1.75	5045.38
2	6000
2.25	7135.24
2.5	8485.28
2.75	10090.8
3	12000
3.25	14270.5
3.5	16970.6
3.75	20181.5
4	24000

Fig. 3.18. Table and graph for exponential growth of bacteria

From the graph, we can see that after 3.25 days, the bacteria count is likely to be between 14 000 and 15 000. The table corroborates this estimate.

Using graphical and tabular representations, students can easily address tasks and questions like the following:

♦ If the swimming-pool bacteria count is 1500 per cubic centimeter on Monday morning at 8 A.M. and the count doubles each day thereafter, what is the bacteria count at noon on the following Tuesday?

♦ If health officials set 200 000 per cubic centimeter as the maximum safe bacteria count for pool water, how long will this pool stay in the safe range?

♦ Calculate each of the following expressions and explain what information the results give about the pool-bacteria situation.

(a) $N(1) - N(0)$

(b) $N(2) - N(1)$

(c) $N(5) - N(4)$

♦ What do these results say about the rate of change in the bacteria count as time passes?

Wage-doubling, bacterial-growth, and even inflation and car-depreciation problems can all be described by exponential functions. We can pose a number of questions related to these situations: "What could I expect to earn on day 22 (given the wage-doubling scheme described earlier) if I were to earn $10 on day 1?" "What if the number of bacteria quadrupled each day rather than doubled or tripled?" "Which car would be more valuable in three years—a $15 000 car that depreciates 30 percent every year or a $12 000 car with a 20 percent depreciation rate?" Using exponential functions of the form $f(x) = Ca^x$ to answer most of these questions amounts to changing the values of C and a. The activity "Exploring the Graphs of Exponential Functions" (Activity 7) engages students in exploring the effects of changing the values of the parameters C and a on the graph of $f(x) = Ca^x$ as well as on the conditions in situations modeled by functions of this form.

The inclusion of exponential functions also affords several other mathematical opportunities. For example, recursion may be introduced and pursued through the recursive definition of an exponential function (see the "dynamical systems" section in chapter 4 of this book). The use of real-world problems, as in the swimming-pool-bacteria situation, can be a natural entry into the consideration of rational and negative values of exponents. The connections among topics should also be part of the students' algebra experiences.

Rational functions (for example, $f(x) = a/x^n$)
A third family of functions that students should consider is rational functions. These functions are useful in considering relationships among variables that involve quotients. Rational functions contrast with linear and exponential functions in that they intrinsically involve restricted domains and vertical asymptotes.

Students' consideration of rational functions may begin with simple quotients. The development of the concept might progress in the following manner:

House-Painting Situation

Earlier we considered several wage-earning schemes. Two were defined by the following pay schedules:

$$D_1(h) = 4.5h - 45$$
$$D_2(h) = 3.5h$$

Determining an algebraic-function rule for the following job offer involves a shift in thinking. Suppose that you have been asked to paint a neighbor's house for $850. Hourly wage is given to be a function of the number of hours worked given that completing the job results in an $850 paycheck. Given this pay scheme, the faster you work, the higher your hourly rate. Similarly, the slower you work, the lower your hourly rate. See the table below.

Number of Hours Worked	Hourly Wage
10	850/10 = 85
11	850/11 = 77.27
12	850/12 = 70.83
13	850/13 = 65.38
14	850/14 = 60.71
⋮	⋮
50	850/50 = 17
51	850/51 = 16.67
52	850/52 = 16.35
53	850/53 = 16.04

The table enables us to address such questions as these: As the number of hours worked to paint the house increases, what happens to the rate at which the hourly wage changes? What would you expect to happen to the hourly wage as the number of hours worked approaches zero? As the number of hours gets very large?

The relation between hours worked and hourly wage can be modeled by a function of the form

$$H(d) = \frac{850}{d},$$

where $d > 0$. This type of function differs from those previously discussed in this chapter. A variable in the denominator yields important function properties. A graph of the function suggests some of those properties (fig. 3.19).

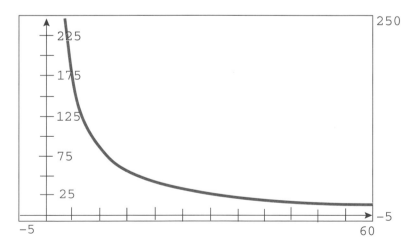

Fig. 3.19. Graphical model of time and hourly wage

The following summary identifies important properties of rational functions of the form $f(x) = a/x$, where a is a constant other than 0.

1. For negative inputs, very small negative inputs (those with very large absolute values) yield outputs close to 0. As negative inputs approach 0, the absolute value of the outputs becomes larger and larger.

2. For positive inputs, inputs very near 0 yield outputs whose absolute values are large. As positive inputs increase, the absolute values of the outputs approach 0.

3. When a is positive, outputs have the same sign as that of the inputs. When a is negative, outputs have a sign opposite that of the inputs.

4. When the input is 0, the output is undefined.

5. Graphs have symmetry about the origin.

6. If

$$f(x) = \frac{a}{x} \quad \text{and} \quad g(x) = \frac{b}{x},$$

where a and b are positive and $a > b$, then the function tables and graphs of these two functions differ in the following ways. In tables, the absolute value of the outputs for $f(x)$ will be larger than those for $g(x)$. The graph of $g(x)$ will be closer to the origin than the graph of $f(x)$.

Polynomial functions of degree 2 and higher

A fourth family of functions is that of polynomial functions of degree 2 and higher. Polynomial functions can be used to model the behavior of variables of interest in many real-world problem settings. For example, second-degree polynomials (quadratics) can be used in studies of projectile motion, the stopping distances of automobiles, and engineering design, such as maximizing the area of enclosed spaces. Unlike linear functions, polynomial functions of degree 2 and higher have varying rates of change. Unlike linear and exponential functions, they sometimes attain local maximum and minimum values. Unlike rational functions, they are continuous over the entire set of real numbers.

Computing tools can aid in the investigation of polynomial functions. Students can start with situations that are well modeled by quadratic functions. They can note the importance of maximum and minimum output values in the context of the situation; they can develop scan-and-zoom or guess-and-test strategies for locating optimum points and intercepts. They can contrast patterns in rates of change of quadratic functions with those of linear functions. Students can explore the effects of changing the values of a, b, and c on the shapes and values of functions of the form $f(x) = ax^2 + bx + c$ for $a \neq 0$.

Consider, as an example, a quadratic function that models the following familiar combinatorial-analysis problem.

Handshake Problem

Determine the number of handshakes if each student in your class shakes hands with every other student once.

Consider this problem by examining simpler cases and looking for patterns. Suppose two students are in the room. How do you respond? Consider three, four, and five students in the room. Try to determine, for each situation, the total number of expected handshakes. Do you suppose that the total number of handshakes will increase at a constant rate as the number of students in the room increases? Consider figure 3.20 and the following reasoning.

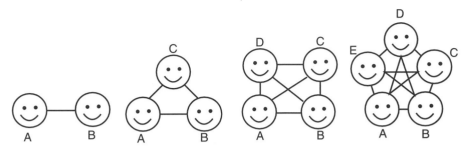

Fig. 3.20. Modeling the handshake problem

Having only two students in the room results in only one handshake. With three students (A, B, and C), there are three handshakes—A and B shake, A and C shake, and B and C shake. For four students (A, B, C, and D), there are six handshakes—A with each of the three others, B with C, B with D, and C with D. Finally, with five students (A, B, C, D, and E), there are ten handshakes—A with each of the other four, B with each of the three still remaining, C with D and E, and D with E.

Notice that even if everyone in the room has already shaken hands with every other person, each additional person who enters adds a certain number of handshakes. When the third person enters, two additional handshakes must occur, since that person must shake hands with each person already in the room. When the fourth person enters, three additional handshakes must occur. When the fifth person enters, four additional handshakes must occur. This pattern is shown in the following table of values.

Number of Students in the Room	Number of Handshakes
2	1
3	3
4	6
5	10

Each person shakes *once* with each other person in the room. If n students are in your class, each person will shake $n - 1$ hands. For instance, with five students in the room, each person will have four others with whom to shake hands. This pattern of handshakes will occur for each of the five students. To determine the total number of possible handshakes, we must avoid counting the same handshake twice. Dividing the total number of handshake patterns, $5 \cdot 4$, by 2, we arrive at the total number of handshakes.

In general, this procedure amounts to the rule

$$H(n) = \frac{n(n-1)}{2}$$

for determining the total number of handshakes when n students are in the room. This algebraic representation can be rewritten in different

Number of Students in the Room	Total Number of Handshakes
1	0
2	1
3	3
4	6
5	10

$1 - 0 = 1$
$3 - 1 = 2$
$6 - 3 = 3$
$10 - 6 = 4$

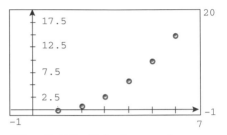

Fig. 3.21. Tabular and graphical representations of the handshake problem

Teaching Matters: The handshake problem suggests some interesting mathematical connections. The number of handshakes in a room with n students is the same as the number of ways to choose two students from n, or

$$\binom{n}{2} = \frac{n!}{(n-2)!\,2!}$$

$$= \frac{n(n-1)}{2} = \frac{1}{2}n(n-1).$$

Ask students to create a different situation and question whose answer is

$$\binom{n}{2}.$$

Ask students to explain why this expression works for their situation.

symbolic forms, each of which describes the same quadratic function:

$$H(n) = \frac{n(n-1)}{2} = \frac{n^2 - n}{2} = \frac{1}{2}n^2 - \frac{1}{2}n$$

The table and graph in figure 3.21 may provide a clearer description of the overall behavior of the handshake function. An analysis of the table and graph reveals that the number of handshakes does not increase at a steady rate as the number of people in the room increases, although the increase in the rate of increase is steady.

Activity 10, "Similarities and Differences in Properties of Different Families of Functions," involves students in synthesizing their experiences with families of functions by comparing and contrasting properties.

CONCLUSION

Nowhere in the teaching of mathematics can the availability of technology have such a profound effect as in the teaching of algebra. The most profound effect on the teaching of algebra will be a radical transformation of its content. In this chapter, we have shown how the content of a technology-intensive algebra can center on the concept of function and on the use of this concept in developing mathematical-modeling ideas. When students are given free access to available computing tools, they can study a powerful array of families of functions. Armed with an intuition about linear, exponential, and rational functions, for example, students can begin to use mathematics in powerful ways to investigate the world around them.

Of course, many other families of functions could be considered. Regardless of the particular families that students might study, the central goals remain the same. Students should be able to use the families of functions as mathematical models in real-world contexts. Students should also develop their own strategies for exploring unfamiliar families of functions in ways that promote their construction of connections among various families of functions and among alternative representations for functions.

A word of caution, however, is in order. Teachers cannot look on a functions approach to algebra as an add-on to traditional algebra curricula. In and of itself, and *without* being supplemented by major ventures into by-hand symbolic manipulation, it is a viable and essential approach to learning about algebraic ideas in a technological world. The issue of the role of symbolic manipulation in this newly defined algebra is developed in chapter 6.

ACTIVITY 3
WINDOW MOLDINGS

The narrow rim of wood that is placed around a window or door is called a *molding*. The amount of molding around any one window or door is related to the size of that particular opening. Start with the example of a rectangular window with molding around all four sides, as shown below. The shaded area represents the molding around the window.

1. Write a function rule that expresses the approximate amount of molding needed as a function of the length (L) and the width (W) of the window:

   $P(L, W) =$ _____

2. Use your rule to answer the following questions about the rectangular window and its molding:

   (a) How much molding would be needed for a rectangular window that measures 4 feet by 2.75 feet?

   (b) What dimensions could a rectangular window have if approximately 28 feet of molding are needed to surround it?

   (c) If the width of a rectangular window remains constant, how will changing the length affect the amount of molding needed to surround it?

   (d) How does function P differ from a function that models the approximate amount of molding needed to surround a hexagonal window, as shown below?

3. In what ways was the model you created a good model for the window-molding situation? In what ways was it not a good model? What else would you need to know to improve your model?

ACTIVITY 4
PET WARDS

A national pet-hotel chain is planning to build wards for a series of franchises. Each unit is a double row of square wards with two-meter-by-two-meter floors as shown in the plan below. Each row always has the same number of wards.

Floor plan

Walls come only in panels that are two meters long. The number of two-meter wall panels needed depends on the number of wards to be included in the unit. Notice that the particular unit shown has length 6 and that this unit has 12 wards and 32 wall panels. Because the chain plans to build many double-row units of different lengths, the manager wants to have a rule relating the number of wards and the number of wall panels.

1. With your partner, create at least one rule with which the manager can determine the number of wall panels on the basis of the number of wards. Explain why your rule is correct.

2. Compare your rule(s) with those created by other pairs of students. Describe the similarities and differences that you observe.

3. How do you know whether your rule(s) and those of other groups are the same? Use your computing tools to show which rules are or are not the same.

Adapted from *Concepts in Algebra: A Technological Approach,* p. 391. Copyright 1995. Janson Publications, Inc., Dedham, Mass.

ACTIVITY 5
FUEL BILLS

The information below comes from one homeowner's fuel bills from August 1992 through March 1993.

Month Ending	Average Daily Cost of Gas	Average Daily Cost of Electricity	Average Monthly Temperature (°F)
21 March 1993	$1.76	$0.73	27
22 February 1993	$1.97	$0.80	27
19 January 1993	$1.94	$0.65	23
21 December 1992	$1.86	$0.86	31
18 November 1992	$1.41	$0.77	44
17 October 1992	$0.78	$0.82	58
16 September 1992	$0.38	$0.78	69
17 August 1992	$0.35	$0.98	70

Month Ending	Average Daily Use of Gas (CCF)	Average Daily Use of Electricity (KWH)	Average Monthly Temperature (°F)
21 March 1993	3.6	6.7	27
22 February 1993	4.0	7.5	27
19 January 1993	3.9	5.7	23
21 December 1992	3.6	7.8	31
18 November 1992	2.5	7.0	44
17 October 1992	1.2	7.7	58
16 September 1992	0.4	6.5	69
17 August 1992	0.4	8.5	70

Five sets of numbers are shown in the preceding charts: average daily cost of gas, average daily cost of electricity, average daily use of gas, average daily use of electricity, and average monthly temperature. Use this information and appropriate computing tools to answer the following questions:

1. Choose two of the five sets of numbers that you think would most likely be related. For example, would average daily cost of electricity be related to average daily use of gas? Would monthly temperature and average daily use of gas be related?

2. Enter the data that you chose in question 1 into your calculator or computer software. Produce a scatterplot of the data. On the basis of the scatterplot, explain what type of function you think best models the relationship between these two variables.

3. Find a function rule that seems to fit the data well. Graph your function over this scatterplot.

4. (*a*) What criteria did you use to determine the model in question 2?

    (*b*) What criteria did you use to determine the function rule in question 3?

5. In what ways is your function a good model? What limitations does your function model have?

6. Since the goal is to predict monthly costs, it seems reasonable to use average monthly cost instead of average daily cost (and average monthly use instead of average daily use). Analyze these data using monthly costs instead of daily costs and monthly use instead of daily use. Discuss how this change affects the function rules you used.

ACTIVITY 6

EXPLORING LINEARITY

You plan to charge $3.75 an hour for weekend yard work and want to calculate—

♦ the charge for time spent on the yard work—2 hours, 3 hours, 4.5 hours, 10 hours, and so on;

♦ the amount of yard-work time needed to reach some earning goal—$129.00 for a portable CD player or $9.95 for a new CD and so on.

1. Complete the following table showing the charges for 1 through 6 hours of yard work at $3.75 an hour.

Time (in hours)	1	2	3	4	5	6
Charge (in dollars)						

2. Make a rough sketch of the graph of the relation between the number of hours worked and the wages earned on the unscaled axes below.

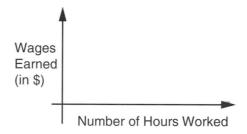

Wages Earned (in $)

Number of Hours Worked

3. Write a function rule describing wages earned as a function w of the number of hours h worked.

4. Using your rule from question 3, write an equation or inequality whose solution provides the answer to the following question: How long must you work to earn at least $25.00? Go ahead and find the answer by producing a table of values or a graph.

Because it is sometimes difficult to get yard workers during the school year, hourly wages are higher. Suppose that during the school year you are hired by a yard-work company at the rate of $4.35 an hour with a $50.00 bonus if you agree to work on a steady basis throughout the year.

5. (*a*) Write a function rule that describes the relation between the number of hours h worked and the amount w you will be paid.

(*b*) Use your rule to predict how much you will be paid for 45 hours of work.

6. With a graphing calculator, generate function tables or graphs to solve this equation derived from a yard-work function rule:

$$4.35h + 50 = 225$$

Or solve this equation using appropriate symbolic-algebra commands for your calculator or computer. Then interpret the answer.

7. (a) Your cousin who lives in another state makes $6.75 an hour for weekend yard work with a bonus of $50.00 for working on a steady basis. Write a function rule describing your cousin's wages earned as a function w of the number of hours h worked.

   (b) The cost of a portable CD player in your cousin's area, however, is $189.00 instead of the $129.00 in your area. Who needs to work fewer hours to purchase a CD player with weekend yard-work income—you or your cousin? What other factors might affect the situation?

   (c) Suppose that your cousin's rate is $7.25 an hour instead of $6.75 an hour. How does this affect your conclusion about the purchase of the CD player?

   (d) Describe two different strategies, primarily using graphs, tables, or rules, for analyzing this situation. Discuss the advantages and disadvantages of each method.

8. Sometimes knowing rates of increase can help in comparing the growth in function values. Assume again that your hourly wage is $3.75 and your cousin's hourly wage is $6.75, that you and your cousin work the same number of hours, that you both work steadily throughout the year, and that she earns $54.00 more than you in a season. Discuss different ways to use the hourly rates to determine how many hours each of you has worked.

ACTIVITY 7

EXPLORING THE GRAPHS OF EXPONENTIAL FUNCTIONS

Recall that for a linear function with rule $f(x) = mx + b$, the values m and b can be used to determine the rate of change in a table or the slope and intercept of a graph. The goal of the following explorations is to help you find clues to understanding the behavior of exponential functions with rules of the form $f(x) = Ca^x$, where $a > 0$ and $C > 0$.

For each of the six exponential functions in the following list, and any other similar functions of your own choosing, use a computer or calculator to make a table of values and a graph. In each table you may choose values of x from -7 through 7 in steps of 1. Use the tabular information as a guide in selecting scales for your graph.

Test functions:

$$f(x) = 0.6^x \qquad\qquad f(x) = 4(0.6)^x$$

$$f(x) = 8(0.6)^x \qquad\qquad f(x) = 2.3^x$$

$$f(x) = 4(2.3)^x \qquad\qquad f(x) = 8(2.3)^x$$

Other functions
 of your choice:

Guidepost questions to assist your investigation of the properties of exponential functions are below.

1. Describe the trends in the relation between inputs and outputs.

   (a) For which of these functions do outputs always increase as inputs increase?

   (b) For which of these functions do outputs always decrease as inputs increase?

   (c) What common features of the rules in (a) make it seem reasonable that these functions should be increasing? What features of the rules in (b) suggest that the functions should be decreasing?

2. Are there any inputs for which at least two of these functions give the same output? If so, can you explain why that would occur?

3. Describe the overall effect of C on the tables of values for functions of the form $f(x) = Ca^x$.

The following questions concern the relation between the exponential function rules $f(x) = Ca^x$ and the patterns of their graphs.

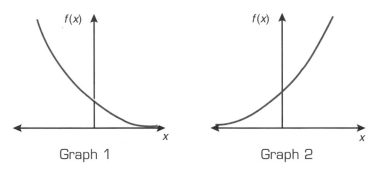

Graph 1   Graph 2

4. (a) Which values of a and C yield graphs like graph 1?

   (b) Which values of a and C yield graphs like graph 2?

5. What is the effect of a on the overall shape of the graph? Why do you think this happens?

6. What is the effect of C on the graph? Explain your answer.

7. Consider this sketch with graphs of three different exponential functions. No grid points or scale tick marks have been provided. Focus only on the overall behavior of the three graphs. Describe ways in which the function rules for each graph are alike. Suggests ways in which they differ.

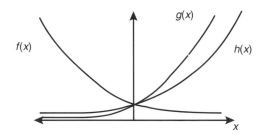

8. Consider the two exponential functions $f(x) = a^x$ and $g(x) = b^x$, where a and b are distinct positive numbers.

   (a) If $a < b$ and each is greater than 1, which function increases faster as x increases?

   (b) If $a > b$ and each is less than 1, which function decreases more rapidly as x increases?

9. Consider two exponential functions of the form $h(x) = Ca^x$ and $j(x) = Da^x$. If $C > D$ and both C and D are positive, what can be said about the relationship between the graphs of the two functions?

ACTIVITY 8

EXPLORING RATIONAL-FUNCTION GRAPHS

The following computer- or calculator-based explorations will assist you in (1) identifying patterns in function tables of (input, output) pairs and (2) describing how these numerical patterns are represented in graphs of functions of the form

$$f(x) = \frac{a}{x},$$

where $a \geq 1$.

For each "test function," produce and inspect a table with inputs between –100 and 100. Special attention should be paid to understanding the behavior of these functions as x approaches 0.

The test functions follow:

(a) $f(x) = \dfrac{1}{x}$ (b) $g(x) = \dfrac{5}{x}$ (c) $h(x) = \dfrac{50}{x}$ (d) $j(x) = \dfrac{100}{x}$

1. Describe what happens to the outputs of each of the four functions as inputs increase from –100 to 100.

2. What do the patterns in the tables suggest about possible maximum (or minimum) outputs for these functions?

3. Are there any inputs for which each function has the same output? If so, what are they?

4. Describe what happens as inputs get very close to 0. What happens when 0 itself is an input?

5. How do differences in a lead to differences in function outputs in each function table?

6. Roughly sketch the general pattern and relative positions of the graphs of the "test functions." Experiment to find graphing windows that nicely display the function graphs.

   (a) What do the patterns in the graphs suggest about possible maximum or minimum outputs for these functions? Describe what happens as x gets large. Tell what happens as x approaches 0.

   (b) Do any of these graphs suggest that outputs change at a constant rate as inputs increase? Explain how the graphs support your conclusions.

   (c) How does the change in the value of the numerator of these functions lead to differences in the function graphs?

Adapted from *Concepts in Algebra: A Technological Approach*, pp. 289–94. Copyright 1995. Janson Publication, Inc., Dedham, Mass.

ACTIVITY 9
CIRCUMFERENCE AND AREA

As a newly hired technical consultant for a chain of hobby stores, you have been asked to use about 25 meters of flexible railroad track for a display in the front window of each store. Your task is to design a track that maximizes the area for display purposes within the perimeter of the track. The general shape of the track will be a figure eight.

By experimenting with the track to determine how much it will bend, you find that the smallest radius the track can attain is about 0.75 meter.

1. Use the following simulation, designed for use with the Geometer's Sketchpad, to determine the maximum area possible. Print out the simulation at the point(s) that you believe maximize the area bounded by the track. The simulation models the railroad-track problem in terms of centimeters rather than meters.

Radius(Circle 1) = 1.40 cm
Circumference(Circle 1) = 8.78 cm
Area(Circle 1) = 6.13 square cm

Radius(Circle 2) = 2.59 cm
Circumference(Circle 2) = 16.27 cm
Area(Circle 2) = 21.07 square cm

Radius(Circle 1)+Radius(Circle 2) = 3.99 cm
Circumference(Circle 1)+Circumference(Circle 2) = 25.05 cm
Area(Circle 1)+Area(Circle 2) = 27.20 square cm

♦ Move the point labeled ← SLIDE → to the left or right.
♦ Observe the changes in values.

2. Does the area of the figure change uniformly as you move the SLIDE point? Why or why not?

3. Construct tables and graphs to describe the relationship between the radius of circle 1 and the sum of the radii of the two circles. Find a rule to describe the relationship. Explain why your rule works.

4. Construct tables and graphs to describe the relationship between the radius of circle 1 and the area of circle 1. Find a rule to describe the relationship. Explain why your rule works.

5. The two functions you found in questions 3 and 4 belong to two different families of functions. Explain why.

The Sketchpad figure shown on the previous page can be constructed using the following steps:

♦ Using the Point tool, choose two points to serve as the centers of Circles 1 and 2.

♦ Highlight these two points and Construct a line Segment between them.

♦ Highlight the segment and Construct a Point on the Object. This point will serve as the SLIDE point.

♦ Highlight Center 1 and then the SLIDE point. Then Construct a Circle by Center + Point.

♦ Highlight Center 2 and then the SLIDE point. Then Construct a Circle by Center + Point.

♦ Highlight Circle 1 and Measure Radius 1, Circumference 1, and Area 1.

♦ Highlight Circle 2 and Measure Radius 2, Circumference 2, and Area 2.

♦ Highlight all measures and enter the Calculate mode.

♦ Calculate Radius 1 + Radius 2, Circumference 1 + Circumference 2, and Area 1 + Area 2.

♦ Label the Center points and SLIDE point as shown and create the indicated caption.

ACTIVITY 10

SIMILARITIES AND DIFFERENCES IN PROPERTIES OF DIFFERENT FAMILIES OF FUNCTIONS

Use the following set of test functions to compare and contrast properties of linear, exponential, quadratic, higher-degree-polynomial, and rational functions. Make notes about the tables and graphs related to your test functions. Record your observations in the chart below. Enlist computing aid as needed when constructing tables of values and function graphs.

Linear	Exponential	Quadratic	Higher-Degree Polynomial	Rational
$f(x) = 2x + 3$	$g(x) = 3^x$	$h(x) = x^2 + 10x + 4$	$k(x) = 0.5x^3 - 6x$	$m(x) = 8/x$
$f(x) = -2x + 3$	$g(x) = (0.3)^x$	$h(x) = -x^2 + 10x + 4$	$k(x) = -0.5x^3 + 6x$	$m(x) = 8/x^2$

SIMILARITIES AND DIFFERENCES IN PROPERTIES OF DIFFERENT FAMILIES OF FUNCTIONS

Summary of Observations

Function Form	Rate of Change[1]	Symmetry Feature[2]	No. of Max/ Min Values[3]	Special Features[4]
$f(x) = mx + b$				
$f(x) = a^x$ (Exponential)				
$f(x) = ax^2 + bx + c$ (Quadratic)				
$f(x) = ax^3 + bx^2 + cx + d$ (Cubic)				
$f(x) = ax^4 + bx^3 + cx^2 + dx + e$ (Quartic)				
$f(x) = \dfrac{a}{x}$				
$f(x) = \dfrac{a}{x^2}$				
$f(x) = \dfrac{a}{x^3}$				
$f(x) = \dfrac{a}{x^4}$				

1. Variable (increasing or decreasing) or constant
2. Symmetry about the vertical axis, about the origin, or about a vertical line that is not the vertical axis
3. None, one, two, and so on
4. Mention features that make this family different from any others in the table.

Adapted from *Concepts in Algebra: A Technological Approach,* pp. 302–3. Copyright 1995. Janson Publication, Inc., Dedham, Mass.

ACTIVITY 11

APPLICATIONS OF POLYNOMIAL FUNCTIONS

The Park and Planning Commission decided to consider three factors when attempting to improve the daily profits at their sports facility:

◆ The number of all-day admission tickets sold

◆ The cost of operating the facility

◆ The price of each all-day admission ticket

After carefully analyzing their operating costs, they found that it would be impossible to cut them further.

Daily Operating Costs

Advertisements	$ 55.00
Employees' pay	$310.00
Heat, lights, taxes, food, rent	$435.00

Knowing that the maximum number of potential patrons is 200, the Park and Planning Commission decided to vary the price of each admission ticket to see what effect this change might have on the number of tickets sold. After much experimentation, they collected the following sales data:

Ticket Price (in $)	Average Number of Tickets Sold
5	158
7	142
9	119
11	97

1. Using this information, suggest the optimal ticket price for all-day admission to the sports facility. If you feel the need for more information, please explain why.

2. Use a graphing tool to find the function rule of best fit for the (price, sales) data. Express the number of tickets sold as a function N of the ticket price. That is, $N(x) = ?$, where x is the price of the ticket. Produce a rough sketch of the (price, sales) data and the fitted function. Do you expect the function to behave like any of the linear, exponential, or rational functions studied so far? If yes, explain your reasoning.

The Park and Planning Commission used an expanded set of (price, sales) data and two basic economic principles,

$$\text{Revenue} = \text{Price charged per item} \times \text{number of items sold}$$

and

$$\text{Profit} = \text{Revenue} - \text{Costs,}$$

to arrive at the following rule for daily profit as a function p of the price of an all-day admission ticket:

$$p(x) = (-8x + 200)x - 800,$$

or

$$p(x) = -8x^2 + 200x - 800.$$

3. Use this rule to generate a function table and graph for ticket prices ranging from $0 to $25 in increments of $1. Record your table and a rough sketch of your graph on a separate piece of paper.

   (a) Describe the trend in profits as prices increase.

   (b) What properties of the relation between price and profit can be best learned from the table? From the graph?

4. What information does the solution to the equation below reveal about the Park and Planning Commission situation?

$$-8x^2 + 200x - 800 = 0$$

Use your table and graph from question 3.

5. (a) The manager estimates that each person spends an average of $3.50 on refreshments. How would you modify the daily profit function,

$$p(x) = x(-8x + 200) - 800 = -8x^2 + 200x - 800,$$

   to reflect the revenue earned from concession sales?

   (b) Compare the function tables and graphs of the original and new profit rules.

6. Taking into account all the foregoing information, recommend the optimal ticket price for all-day admission to the sports facility. Support your recommendation with function tables and graphs.

CHAPTER 4
EXTENDING A FUNCTIONS APPROACH

With their major algebraic experiences rooted in work with functions and mathematical modeling, students develop intuitions that can be useful in investigating more-complex applications of functions. The concept of function can easily be extended to the simultaneous consideration of several functions or to the consideration of functions of two variables. The stretch to functions of several variables is not difficult as long as a graphical representation is not required.

Within a functions approach, the traditional focus on a system of equations can shift to the idea of a system of functions. Students can begin to view an equation such as $7.5a + 3.75c = 9000$ as a statement about the function f of two variables with rule $f(a, c) = 7.5a + 3.75c$. From a functions perspective, the study of systems becomes the study of a dynamic process. The concept of function can also extend naturally to "functions of functions," and students can study more complicated relationships by viewing them in terms of composite functions. Finally, viewing algebra from a functions perspective opens new mathematical worlds to high school students. Students can begin to understand some of the issues related to dynamical systems, the mathematics of change, and the mathematics that describes so many newly discovered patterns and phenomena. This chapter introduces some of the ways that a functions approach can extend students' mathematical perspectives.

SYSTEMS INVOLVING SEVERAL FUNCTIONS OF ONE VARIABLE

As was noted in the previous chapter, many real-world situations suggest single relationships that are well modeled by a single linear, exponential, or rational function. A large number of other situations, however, are better described using not one but several functions, each of which characterizes a different aspect of the situation.

Economic situations offer some of the most common examples of using several functions to describe a situation. For example, in chapter 3 we considered the Park and Planning Commission situation by investigating four functions of the price t of an admission ticket:

♦ A demand function N with rule $N(t) = -8t + 200$

♦ A revenue function R with rule $R(t) = t(-8t + 200)$

♦ A cost function C with rule $C(t) = 800$

♦ A profit function P with rule $P(t) = t(-8t + 200) - 800$

When several functions describe a single situation and depend on a single variable, it is often convenient to use simultaneous tables and graphs to analyze the situation. For the park-and-planning situation, an examination of the charts in figures 4.1, 4.2, and 4.3 and the graphs in figure 4.4 helps in a search for a maximum profit and also suggests a number of global features of the functions.

For example, class discussion could center on the zeros of each function:

♦ How many zeros does the revenue function have? How do we know? Why is this conclusion reasonable? Why does it make sense in terms of the park-and-planning situation?

Start = 0 Increment = 2	$N(t) = 200 - 8t$	$R(t) = t(200 - 8t)$	$C(t) = 800$	$P(t) = t(200 - 8t) - 800$
0	200	0	800	−800
2	184	368	800	−432
4	168	672	800	−128
6	152	912	800	112
8	136	1088	800	288
10	120	1200	800	400
12	104	1248	800	448
14	88	1232	800	432
16	72	1152	800	352
18	56	1008	800	208
20	40	800	800	0
22	24	528	800	−272
24	8	192	800	−608

Fig. 4.1. Demand, revenue, cost, and profit function values (for admission-ticket prices from $0 to $24 in increments of $2) for the Park and Planning Commission situation

Start = 12 Increment = 0.2	$N(t) = 200 - 8t$	$R(t) = t(200 - 8t)$	$C(t) = 800$	$P(t) = t(200 - 8t) - 800$
12.0	104.0	1248.00	800	448.00
12.2	102.4	1249.28	800	449.28
12.4	100.8	1249.92	800	449.92
12.6	99.2	1249.92	800	449.92
12.8	97.6	1249.28	800	449.28
13.0	96.0	1248.00	800	448.00
13.2	94.4	1246.08	800	446.08

Fig. 4.2. Demand, revenue, cost, and profit function values (for admission-ticket prices from $12.00 to $13.20 in increments of $0.20) for the Park and Planning Commission situation

Start = 12.4 Increment = 0.05	$N(t) = 200 - 8t$	$R(t) = t(200 - 8t)$	$C(t) = 800$	$P(t) = t(200 - 8t) - 800$
12.40	100.8	1249.92	800	449.92
12.45	100.4	1249.98	800	449.98
12.50	100.0	1250.00	800	450.00
12.55	99.6	1249.98	800	449.98
12.60	99.2	1249.92	800	449.92

Fig. 4.3. Demand, revenue, cost, and profit function values (for admission-ticket prices from $12.40 to $12.60 in increments of $0.05) for the Park and Planning Commission situation

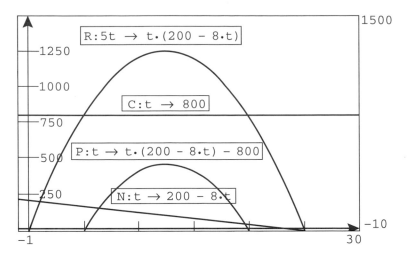

Fig. 4.4. Graphs of demand, revenue, cost, and profit functions for the Park and Planning Commission situation

♦ Does the demand function have a zero? Where is it? What does it mean in terms of the park-and-planning situation?

♦ How many zeros does the profit function have? How do we know? Why is this conclusion reasonable in light of the park-and-planning situation?

Similarly, a consideration of rates of change could precipitate some interesting class discussion:

♦ On the basis of the charts in figures 4.1, 4.2, and 4.3, determine which function seems to have a constant rate of change. Why does it make sense that functions like these have constant rates of change? Will these types of functions always have constant rates of change? Explain your conclusion.

♦ Which functions do not have a constant rate of change? Does any pattern exist in the rates at which these functions increase or decrease? Does this make sense in terms of the park-and-planning situation? Explain.

A variety of other questions related to the more global features of the functions may arise in class discussions.

To focus the students' attention on the contextual meaning of function shapes and values, teachers can pose specific questions about the functions and the situation. For example, students might investigate the following questions:

♦ For which admission-ticket prices does the Park and Planning Commission break even? Explain your reasoning.

♦ What is the highest price that can be charged without losing all paying customers? Explain how you can tell.

♦ What would you advise the Park and Planning Commission to charge for an admission ticket? Explain why.

Each question could be answered by using spreadsheets, graphs, tables, or the results from a symbolic-manipulation program. Figure 4.5 shows the numerical results when a symbolic manipulation program is asked to solve P(t) = 0 for t and N(t) = 0 for t after P and N have been defined.

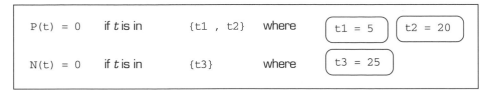

```
P(t) = 0    if t is in    {t1 , t2}   where    [ t1 = 5 ]   [ t2 = 20 ]

N(t) = 0    if t is in    {t3}        where    [ t3 = 25 ]
```

Fig. 4.5. The results obtained when the symbolic-manipulation program Calculus T/L II is asked to solve P(t) = 0 for t and N(t) = 0 for t after the user has defined P and N

The simultaneous display of tables and graphs makes it feasible to consider the four functions as an interrelated system of functions. It also engenders "what if" questions about the relationships. For example, these questions might be asked:

♦ The cost function in this situation is constant.

   (a) What effect does a constant cost function have on the relationship between the revenue and the profit functions?

   (b) What would happen to the relationship between revenue and profit if the cost function were not constant but still linear?

♦ The maximum revenue and the maximum profit both seem to occur when the ticket price is about $12.50.

   (a) Is there enough evidence in the tables or the graphs to establish $12.50 as the optimal ticket price, or is further zooming in needed?

   (b) Under what conditions will the maximum revenue and the maximum profit occur for different ticket prices?

SYSTEMS INVOLVING FUNCTIONS OF SEVERAL VARIABLES

At times, relationships are not expressible as functions of a single variable. Examples of functions of several variables occur in a wide range of settings: the revenue from ticket sales may be a function of both ticket price and the amount of money spent in advertising; the volume of a cylindrical can is a function of both the height of the can and its radius; and the number of calories consumed in a particular meal is a function of the amounts of each food consumed.

Problems that involve functions of several variables often have as their premise a need to find an appropriate balance in the values of the variables. We may want to find the combination of ticket price and advertising investment that will maximize profit, the height and radius of a tin can that will minimize the cost of production, or the amounts of each food that must be eaten to reach a particular goal for calorie consumption.

The first functions of several variables that we will discuss describe the nutritional content of the foods we eat. First, we will consider information from the 1980 *Recommended Daily Dietary Allowances* of the National Research Council (see fig. 4.6).

◆　　◆　　◆　　◆　　◆　　◆　　◆　　◆

	Male 11–14 yrs	Male 15–18 yrs	Female 11–14 yrs	Female 15–18 yrs
Vitamin C (mg)	50	60	50	60
Calcium (mg)	1200	1200	1200	1200
Iron (mg)	18	18	18	18
Protein (g)	45	56	46	46
Sodium (mg)	900–2700	900–2700	900–2700	900–2700
Calories	2700	2800	2200	2100

Fig. 4.6. Daily nutritional needs for teenaged males and females. Excerpted from Recommended Daily Dietary Allowances (National Research Council 1980)

These excerpts give information about the amount of daily caloric intake, protein, and the amount of each of four minerals and vitamins recommended for teenagers of typical height and weight. The actual recommended amounts vary and depend on a particular person's body type, height, weight, and level of activity.

In planning a diet, or in figuring out the nutritional value of our current diet, it is helpful to know the nutritional content of the foods we eat. Knowing the recommended daily allowances is not sufficient information. We also need to know the quantity of each vitamin or mineral in a serving of each food. The chart in figure 4.7 gives vitamin, mineral, protein, and calorie information for a range of foods that might be eaten in a single day.

Consider the nutritional value of a ham-and-cheese sandwich. For this particular sandwich, we will use a quarter-pound of cooked ham, one slice of Swiss cheese, two slices of whole-wheat bread, and one-half tablespoon of mustard. The number of calories in the sandwich is a function of the amount of each ingredient. The total number of calories in the sandwich is a function C of the amount of each ingredient—ham (h), cheese (c), bread (b), and mustard (m)—with the following rule:

$$C(h, c, b, m) = 424h + 75c + 61b + 14m$$

In this example, the total number of calories is $C(1, 1, 2, 1/2)$. Since $C(1, 1, 2, 1/2) = 424(1) + 75(1) + 61(2) + 14(1/2) = 628$, the total number of calories in this sandwich is 628.

"Using Functions of Many Variables to Analyze Dietary Choices" (Activity 12) gives students an opportunity to learn about functions and about the nutritional value of the foods they eat. **Students will need copies of figure 4.7 to complete the activity.**

Just as two-dimensional graphical representations can highlight the more global features of functions of one variable, three-dimensional graphical representations are helpful in understanding features of functions of two variables.

Consider creating three cups of a snack mix that comprises chocolate chips and raisins. Let c represent the number of cups of chocolate chips in the

Item	Amount	Vitamin C (mg)	Calcium (mg)	Iron (mg)	Protein (g)	Sodium (mg)	Calories
Banana	1	10	10	0.8	1.3	1	101
Bread (rye)	1 slice	0	19	0.4	2.3	139	61
Bread (whole wheat)	1 slice	T	25	0.8	2.6	132	61
Broccoli	1 cup	135	132	1.2	4.7	15	39
Butter	1 tsp	0	1	0.0	0.0	65	48
Carrots	1 cup	9	50	0.9	1.4	50	47
Cheese (Swiss)	1 slice	0	186	0.2	5.5	245	75
Chocolate chips	1/6 cup	T	9	0.7	1.2	1	144
Egg (scrambled)	1	–	45	1.0	6.3	144	97
Ham (cooked)	0.25 lb	0	11.3	3.4	26.0	64	424
Hamburger	0.25 lb	0	14	4.0	31.1	4	248
Lettuce	1 cup	16	70	4.0	2.4	18	28
Mayonnaise	1 Tbsp	0	3	0.1	0.2	84	101
Milk (low fat)	1 cup	2	352	0.1	10.3	150	145
Milk (skim)	1 cup	2	296	0.1	8.8	127	88
Mustard	1 Tbsp	0	19	0.3	0.9	196	14
Peanuts (salted)	1/6 cup	0	20	0.6	7.0	114	159
Pork chop (baked)	0.25 lb	0	15	4.3	33.4	82	288
Potato (baked)	1	31	14	1.1	4.0	6	145
Raisins	1 cup	1	90	5.1	3.6	39	419
Raspberries	1 cup	24	40	1.2	2.0	1	98
Roll (hard)	1	T	24	0.4	4.9	313	156
Steak (T-bone)	0.25 lb	0	9	3.0	22.1	54	537

Fig. 4.7. Vitamin, mineral, protein, and calorie information for a range of foods that might be eaten in a single day (source: Silverman, H. M., J. A. Roman, and G. Elmer. The Vitamin Book: A No-nonsense Guide [New York: Bantam Books, 1985])

mix, and let r represent the number of cups of raisins in the mix. We can analyze the nutritional content of various balances of the snack mix.

First, look at the caloric content of the snack mix. The number of calories in the mix is a function N of the amount of each ingredient. Using the facts that one-sixth cup of chocolate chips has 144 calories (so a cup has 864 calories) and a cup of raisins has 419 calories, we can see that the function rule for N is $N(c, r) = 864c + 419r$. The chart in figure 4.8 gives values of N for combinations using up to three cups of raisins and three cups of chocolate chips.

A number of interesting patterns appear in this chart. For example, look at the numbers across the second row:

140   428   716   1004   1292   1580   1868   2156   2444   2732

Amount of Chocolate Chips

Amount of Raisins	0 cup	$\frac{1}{3}$ cup	$\frac{2}{3}$ cup	1 cup	$1\frac{1}{3}$ cups	$1\frac{2}{3}$ cups	2 cups	$2\frac{1}{3}$ cups	$2\frac{2}{3}$ cups	3 cups
0 cup	0	288	576	864	1152	1440	1728	2016	2304	2592
$\frac{1}{3}$ cup	140	428	716	1004	1292	1580	1868	2156	2444	2732
$\frac{2}{3}$ cup	279	567	855	1143	1431	1719	2007	2295	2583	2871
1 cup	419	707	995	1283	1571	1859	2147	2435	2723	3011
$1\frac{1}{3}$ cups	559	847	1135	1423	1711	1999	2287	2575	2863	3151
$1\frac{2}{3}$ cups	698	986	1274	1562	1850	2138	2426	2714	3002	3290
2 cups	838	1126	1414	1702	1990	2278	2566	2854	3142	3430
$2\frac{1}{3}$ cups	978	1266	1554	1842	2130	2418	2706	2994	3282	3570
$2\frac{2}{3}$ cups	1117	1405	1693	1981	2269	2557	2845	3133	3421	3709
3 cups	1257	1545	1833	2121	2409	2697	2985	3273	3561	3849

Fig. 4.8. *Number of calories in the snack mix, with the mixtures that amount to no more than three cups displayed in boxes*

Now subtract adjacent entries:

$$2732 - 2444 = 288$$
$$2444 - 2156 = 288$$
$$2156 - 1868 = 288$$
$$1868 - 1580 = 288$$
$$1580 - 1292 = 288$$
$$1292 - 1004 = 288$$
$$1004 - 716 = 288$$
$$716 - 428 = 288$$
$$428 - 140 = 288$$

In fact, the difference between horizontally adjacent numbers is 288, no matter which row of numbers is chosen. It would be interesting for students to discuss the relationship between 288 and the nature of the snack mix.

A graphical representation (fig. 4.9) of the function $N(c, r) = 864c + 419r$ highlights the rate of change in still a different way. Class discussion could focus on the meaning of the darkened lines and on the relationship among the slopes of these lines. Possible questions for small-group or whole-class discussion might include the following:

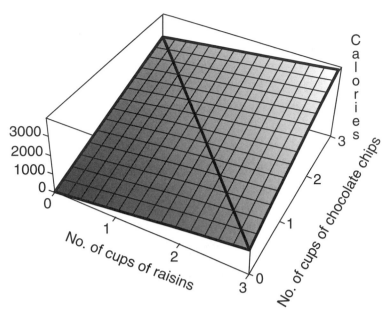

Fig. 4.9. A graph of the function N with rule N(c, r) = 864c + 419r produced using the software package Mathematica

♦ What does each of the five darkened line segments represent in terms of the snack mix?

♦ Which of these line segments has greatest slope?

♦ Which of these line segments has least slope?

♦ Interpret the slopes of the line segments in terms of the snack mix.

Similar three-dimensional graphs could be used as a catalyst for discussing the marginal benefit of raisins or chocolate chips on sodium intake or on the amount of vitamin C in the snack mix consumed.

A system of equations or inequalities related to functions of several variables could be used to address questions such as the following:

♦ Which, if any, three-cup combinations of snack mix containing some raisins and some chocolate chips have fewer than 30 mg of sodium and fewer than 2000 calories?

In this example, note that the amount of sodium in the snack mix is a function S of the amount of each ingredient, with rule $S(c, r) = 6c + 39r$. Answering this question amounts to solving the system

$$c > 0,$$
$$r > 0,$$
$$S(c, r) < 30,$$

and

$$N(c, r) < 2000.$$

Or, equivalently,

$$c > 0,$$
$$r > 0,$$
$$6c + 39r < 30,$$

and

$$864c + 419r < 2000.$$

Try This: A spreadsheet is an excellent tool for analyzing functions of several variables. A practical project using different functions of several variables is the nutritional analysis of diets and menus. If students are familiar with spreadsheets, have them construct a spreadsheet that would make it easy to calculate the number of calories and the amount of protein or iron for a given diet. If they are not familiar with spreadsheets, construct a template for them. This is an excellent project to conduct in conjunction with a health or nutrition class.

It is interesting to pursue graphical and numerical solutions to questions like this. How could we find a solution to this system using three-dimensional graphs of N and S? How could we find a solution using tables of values for N and S as functions of c and r? Although such graphical and numerical solutions may take more time than solutions that rely on traditional by-hand symbolic-manipulation techniques, unlike symbolic-manipulation solutions, they keep the solver's attention on the function values of N and S throughout.

COMPOSITE FUNCTIONS

Sometimes it is useful to consider composite functions, or functions that act on the output from other functions. A composite function arises naturally in the context of the following situation.

> **Fund-Raising Situation**
>
> As a fund-raising activity each summer, a school band runs the refreshment stands on weekends at the regional amusement park. Local businesses contribute all the refreshments and other supplies, but the band must pay $110.00 a week to rent the stands for the twelve-week season. In the course of analyzing the situation, band members figured out that the number of admission tickets sold for the season would be a function of the price of an admission ticket with the rule $N(t) = -80t + 1250$. From past experience, the band also knew that (1) the profit will be a function P of the number of admission tickets sold and (2) on average, refreshment-stand revenue will be approximately $4.75 for each paid admission.

After a complete analysis of the situation, the band members compiled function rules for number of tickets sold and for seasonal profit:

$$N(t) = -80t + 1250,$$

where t is the price of an admission ticket

$$P(x) = 4.75x - 1320,$$

where x is the number of paid admissions

The following input-output diagram further illustrates the relationships among ticket price, number of tickets sold, and seasonal profit for the weekend refreshment stand.

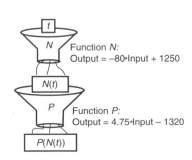

Notice that the input for function P is itself a function. In symbolic notation, $P \circ N$ is a composite function whose output values are determined by first applying the function N to the input values and then applying the function P to the output values from N. Designate the output value for an input value of a as $P(N(a))$. Activity 13, "Refreshment-Stand Profits," at the end of this chapter gives students the chance to explore this composite-function situation.

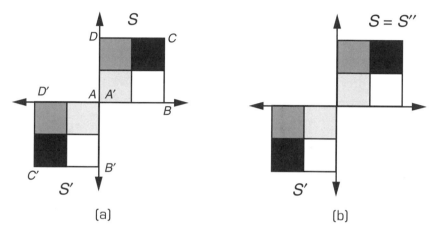

Additional examples of composite functions can be found in chapter 5, "Matrices." For example, consider a function f that takes as input a point, rotates the point 90 degrees clockwise about the origin, and reflects it over the vertical axis. If the inputs are the points of a unit square in standard position, then the outputs would be the points of a unit square with vertices at $(0, 0)$, $(0, -1)$, $(-1, -1)$, and $(-1, 0)$ as shown in figure 4.10a.

We can see what effect the function f has on the unit square S. What happens if f is applied to S', where $S' = f(S)$? The result is shown in figure 4.10b.

Fig. 4.10. The graph of f(S) = S′ (where f is a function that takes as input a point, rotates the point 90 degrees clockwise about the origin, and reflects it over the vertical axis) and f(S′) = S″

We can see from the figure that $f(S') = S$. In other words, $f(f(S)) = S$.

The concept of a composite function is essential to the study of geometric transformations like the reflections and rotations described in this section. Students who study compositions of different kinds of reflections and rotations will find a myriad of patterns and connections.

DYNAMICAL SYSTEMS

The process of applying a function to some initial input, applying the same function to the first output, and applying the same function to each successive output is called *iterating the function*. In recent years, mathematicians have begun to focus on the mathematics behind iteration, and the field of mathematics called *dynamical systems* has flourished. This field has a special vocabulary. When a function undergoes iteration, the initial input is called the *seed* and the sequence of outputs created by iterating the function is called an *orbit*. In dynamical systems, the behavior of orbits created by iterating families of functions is studied. Many interesting mathematical questions arise by iterating basic functions like the linear and the quadratic functions.

If the inputs and outputs for a function are numbers, technology can be used in a variety of ways to generate orbits. Shown below are three different ways to calculate the orbits created by iterating the function $f(x) = x^2 + c$. All three programs allow the user to input a seed x_0 and a value for the parameter c.

Try This: Have students construct a grid and a unit cardboard square with its quadrants shaded similarly on both sides to answer the following question:

Let S be the unit square with vertices (0, 0), (0, −1), (−1, −1), and (−1, 0). Let f be a function that takes as input a unit square, rotates it 90 degrees clockwise about the origin, and reflects it over the vertical axis. Designate f(f(S)) as $f^2(S)$, f(f(f(S))) as $f^3(S)$, and so on. For what values of n is it true that $f^n(S) = S$?

A program for the TI-81 or TI-82 programmable graphing calculator is given in figure 4.11, along with the output for an input of $X = 2$, $C = -3.5$, and $N = 5$.

```
Prgm 1: ITERATE
: Disp "X"
: Input X
: Disp "C"
: Input C
: Disp "N"
: Input N
: 1 -> I
: Lbl 1
: (X^2+C) -> X
: Disp X
: I + 1 -> I
: If I<N
: Goto 1
: Stop
```

Output

```
Prgm1
X
?2
C
?-3.5
N
?5
                        .5
                     -3.25
                    7.0625
               46.37890625
```

Fig. 4.11. TI-81 or TI-82 program for calculating orbits

Figure 4.12 shows a program using the programming language True BASIC.

```
Input prompt "c?"
Input prompt "x0?":x
Input prompt "How many iterations?":n
for i = 1 to n
        print x
        let x=x^2+c
next i
end
```

Fig. 4.12. True BASIC program for calculating orbits

A third approach uses a spreadsheet program. After the seed value of 0.5 and the parameter value of –3.5 are entered into cells B3 and B4, the output is calculated for cell D3 with the formula B3*B3 + B4. The calculations continue with the entry in cell D(k) resulting from the formula (D(k – 1) * D(k – 1) + B4). The resulting table appears in figure 4.13.

A	B	C	D
	INPUTS		OUTPUTS
SEED	0.5		−3.25000E+000
PARAMETER	−3.5		7.06250E+000
			4.63789E+001
			2.14750E+003
			4.61177E+006
			2.12684E+013
			4.52344E+026
			2.04615E+053
			4.18673E+106
			1.75287E+213

Fig. 4.13. Spreadsheet illustrating the iteration of the function $f(x) = x^2 + c$

Analyzing how the orbits behave for different values of c and different seeds is an activity that has generated many new and important results in the past fifteen years, not the least of which is the discovery of the Mandelbrot set (fig. 4.14), one of the most complex mathematical objects known.

Actually, many functions can be iterated using only a scientific calculator. For example, to iterate $f(x) = x^2$, enter a seed, say 3, and keep pressing the x^2 key. "Exploring Function Orbits" (Activity 14) shows some interesting patterns arising from the iteration of functions.

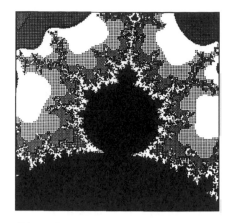

Fig. 4.14. A Mandelbrot set

Using Iterated Linear Functions as Models for Growth and Decay

Iteration can be used to build mathematical models of a variety of dynamic processes. Some examples follow.

Bank balance

Bankers use iteration to calculate the current balance of their customers' accounts. Suppose you invest \$2000 in a savings account that pays 12 percent annual interest compounded monthly.

After one month, you would have

$$2000 + \frac{0.12}{12} \cdot 2000 = (1.01)\, 2000 = 2020$$

dollars. After two months, you would have

$$2020 + (0.01)2020 = (1.01)2020 = 2040.20$$

dollars. After three months, you would have

$$2040.20 + (0.01)2040.20 = (1.01)2040.20 = 2060.60$$

dollars. Notice that the amount for the next month will always be 1.01 times the balance for the current month. Letting x_k denote the balance at the end of the kth month,

$$x_{k+1} = (1.01)x_k.$$

Annuity

Another common financial account is an annuity, which accumulates interest like a savings account but into which regular monthly deposits are made. For example, consider the savings account just described and assume that regular deposits of \$100 a month were made.

If x_k denotes the current balance, then

$$x_{k+1} = (1.01)x_k + 100$$

will denote the balance at the beginning of the next time period. Notice that this equation assumes the payment is deposited right at the beginning of the time period. The balance in this account can be calculated by iterating $f(x) = (1.01)x + 100$, starting with a seed of 2000.

Loan

A loan is a third type of account and is very similar to an annuity, except that now you owe the bank money and your initial balance is negative. For example, suppose you borrowed \$5000 from a bank that charged 12 percent interest and you made monthly payments of \$600 a month.

Teaching Matters: After students have completed activity 14, an interesting class discussion could arise around the iteration of family of functions

$$f(x) = \frac{x + \dfrac{a}{x}}{2}.$$

It turns out that for $a > 0$, iterating this function will produce orbits that tend to either

$$\sqrt{a} \quad or \quad -\sqrt{a}.$$

This is called the Babylonian method of finding square roots and has been known since ancient times. After students have discovered this relationship, have them use the Babylonian method to find approximations for

$$\sqrt{17} \quad and \quad \sqrt{12324}.$$

Try This: Have students work in small groups to come up with ways to solve these problems.

1. A credit-card company offers to lend you money at a rate of 1.5 percent a month. If you borrow $1000 and if you want to pay off the loan within a year, what is the minimum monthly payment you can make?

2. A financial company offers to lend you $16 500 for a five-year period. The company charges an annualized rate of 8.5 percent a year. What is the minimum amount you would have to pay each month?

3. You have $5000 in a bank account and you add $150 a month. If the bank pays at an annualized rate of 4 percent a year, how much will you have at the end of three years?

To calculate the amount still owed at the end of the kth month, start with a seed of –5000 and create an orbit by iterating

$$x_{k+1} = (1.01)x_k + 600.$$

To figure out how long it would take to pay off the loan, keep iterating until x_k becomes nonnegative.

In general, the value of an account with a balance of x_k at the end of one month into which are placed monthly payments of P and that pays r percent each time period will equal $f(x_k)$ at the next time period, where

$$f(x) = (1 + r)x + P.$$

The value of the account after k time periods can be found by iterating f k times or by calculating the kth iterate of f, which is denoted by $f^{[k]}(x_0)$.

In the preceding examples, iteration of linear functions was used to model quantities that were growing. They can also be used to model quantities that are decreasing. For example, people who own swimming pools have to add chlorine to keep down the amount of bacteria. Chlorine evaporates in such a way that each day, 5 percent of the chlorine disappears. This rate means that if the amount of chlorine is currently ten units, then the amount one day later will equal

$$10 - (0.05)10 = (1 - 0.05)10 = (0.95)10 = 9.5 \text{ units,}$$

the amount two days later would be

$$(0.95)9.5 = 9.025 \text{ units,}$$

and the amount three days later will equal

$$(0.95)9.025 = 8.57375 \text{ units.}$$

Note that to model the concentration of chlorine, the function $f(x) = (0.95)x$ has been iterated. The amount of chlorine becomes less and less, and the chance of having harmful bacteria becomes greater and greater. Most pool owners do not add just a single dose when they fill their pool; rather, they add a smaller dose at regular intervals. For example, suppose the owner added two units of chlorine every day. What would the amount of chlorine in the pool be in the long run? This situation can be modeled by iterating the function $f(x) = 0.95x + 2$, starting with a seed of 10. Using technology to calculate the orbit gives the following results:

$$10, 11.5, 12.925, \ldots, 40, \ldots.$$

If the owner started with 30 units, the result would be

$$30, 30.5, 30.975, \ldots, 40, \ldots.$$

If the owner started with 40 units, the result would be

$$40, 40, 40, \ldots, 40, \ldots.$$

If the owner started with 100 units, the result would be

$$97, 94.15, 91.4425, 88.870\,375, \ldots, 40, \ldots.$$

Notice that the starting value seems to make no difference; the orbits always seem to approach the value 40.

The preceding examples illustrate the different ways that the orbits created by iterating a linear function can behave: they either grow larger and larger or they seem to approach some limiting value. This behavior can be explained by a little analysis. Some useful facts about the iteration of linear functions in general follow.

Fact 1. If $a \neq 1$, then there is some x_1 such that $f(x_1) = x_1$. We call x_1 a *fixed point.* To find x_1, all that is needed is to solve the equation

$$x_1 = ax_1 + b,$$

getting

$$x_1 = \frac{b}{1-a}.$$

Fact 2. If $f(x) = ax + b$ has a fixed point, then f can always be expressed in the form

$$f(x) = a(x - x_1) + x_1,$$

where

$$x_1 = \frac{b}{1-a}.$$

The equation form in fact 2 is called the *fixed-point form* of a linear equation. It furnishes an interesting geometric interpretation of what happens when a linear function is evaluated. Think of f as a transformation that maps points on a number line to other points on a number line or, put simply, f moves x to $f(x)$.

Fact 3. Suppose $f(x) = ax + b$ ($a \neq 1$) has a fixed point $x_1 = b/(1 - a)$. If $a < 1$, then the iterates of f move closer to x_1 by a factor of a and the fixed point is said to be *attracting.* If $a > 1$, then the iterates move away from x_1 by a factor of a and the fixed point is said to be *repelling.*

Suppose a prospective homeowner borrows money at 9 percent annualized interest. Then $f(x) = (1 + 0.09/12)x + P$ is used to calculate the amount of the mortgage when payments of P dollars are made. Here, $f(x) = 1.0075x + P$. The fixed point is

$$x_1 = \frac{b}{1-a} = \frac{P}{1-1.0075} = \frac{P}{-0.0075}.$$

What does the fixed point tell about a particular loan? If the payments on this 9 percent annualized-interest loan are $675, for example, then $f(x) = 1.0075x + 675$. The fixed point is $675/-0.0075 = -90\ 000$, and each payment is 1.0075 times as far from $-90\ 000$ as before the payment. In other words, suppose a person borrows $40 000 at 9 percent annualized interest. Each payment moves the amount owed away from $-90\ 000$. The first payment of $675 moves the amount owed away from $-90\ 000$ by a factor of 1.0075. In other words, the amount owed before the payment was $40 000, which is $50 000 less than $90 000. After the payment, the amount owed is $1.0075(50\ 000) = 50\ 375$ away from 90 000. In other words, $39 625 ($90\ 000 - 50\ 375 = 39\ 625$) is still owed on the principal.

These relationships are useful in a number of different ways. First, if we know what we can afford to pay each month, we can put an upper bound on what we can afford to borrow. For example, if we can afford to pay no more than $600 a month, then we should not even consider bor-

x	$P = -0.0075x$
0	0
–10 000	75
–20 000	150
–30 000	225
–40 000	300
–50 000	375
–60 000	450
–70 000	525
–80 000	600
–90 000	675
–100 000	750

Fig. 4.15. Debt incurred and loan payments

rowing more than $80 000 (see fig. 4.15). The upper bound is not what we can afford to borrow, however, since paying $600 a month on an $80 000 loan means that we would never pay off even one cent of the principal! We say that $80 000 is a fixed point for a loan on which we are paying $600. In most mortgages, the payments are such that the amount owed gets reduced. In other words, the amount owed gets repelled from the fixed point. Since the amount owed moves away from the fixed point in a geometric progression, it moves very slowly at first and then picks up speed. This explains the frustration a borrower feels when paying on a mortgage; it takes so long before the principal begins to get significantly smaller.

Technology now makes accessible to beginning algebra students a variety of other applications of iterating functions. For example, iterating functions can be used to study the maximum sustainable population for a given environment. Iterating functions and fractals help explain chaotic behavior.

CONCLUSION

With a working knowledge of functions, students can expand their work with functions to other areas of mathematics. Early in their formal algebra experience, students can begin to see systems from a functions perspective, can develop the sophistication needed to study functions of functions, and can glimpse the world of dynamical systems. With a functions approach to algebra, new worlds of geometric transformations and dynamical systems open readily to technology-equipped students.

ACTIVITY 12

USING FUNCTIONS OF MANY VARIABLES TO ANALYZE DIETARY CHOICES

The *New American Eating Guide,* developed by the Center for Science in the Public Interest suggests that people distribute their dietary choices each day according to the following guidelines:

> 4 or more servings of beans, grains, and nuts
>
> 4 or more servings of fruits and vegetables
>
> 2 servings of milk products
>
> 2 servings of poultry, fish, meat, and eggs

A sample menu that one thirteen-year-old male followed on one day is given below.

Breakfast

3 slices buttered rye toast
2 scrambled eggs
1 cup skim milk

Lunch

2 hamburger sandwiches (each in a hamburger roll with mustard)
1 large lettuce salad (2 cups lettuce with 2 tablespoons mayonnaise)
2 cups skim milk

Dinner

1 baked pork chop
1 baked potato with butter
2 servings broccoli
1 cup lettuce salad with 1 tablespoon mayonnaise
1 cup skim milk

The total number of calories he consumed is a function of the number of servings for each food. The information in the chart from *The Vitamin Book* can be used to generate a function rule C_B that describes the total number of calories in his breakfast:

$$C_B(r, b, e, m) = 61r + 48b + 97e + 88m$$

You can calculate that 609 calories are in this breakfast, since

$$C_B(3, 3, 2, 1) = 61(3) + 48(3) + 97(2) + 88(1) = 609.$$

In a similar manner, you could calculate the number of calories in the other two meals or the amount of protein, calcium, sodium, iron, or vitamin C in the day's menu. A spreadsheet was used to generate the totals appearing in the chart on the next page. The two left columns show the total number of calories and the amounts of certain vitamins and minerals in the menu under consideration. The two right columns show the amounts of each of these measures recommended in the daily diet of a thirteen-year-old male.

Total Calories	2459	Desired Calories	2700
Total Protein	159.3	Required Protein (g)	45
Total Vitamin C	341	Required Vitamin C (mg)	50
Total Calcium	1834	Required Calcium (mg)	1200
Total Iron	27	Required Iron (mg)	18
Total Sodium	2656	Required Sodium (mg)	900–2700

1. Analyze the given menu to determine how well it fits the guidelines.

2. Construct a menu that corresponds to the *New American Eating Guide.* If necessary, do more library research to find the nutritional information about the foods you include.

3. Construct a spreadsheet to calculate the total calories, protein, calcium, vitamin C, iron, and sodium in a given menu.

4. Use your spreadsheet to calculate the total calories, protein, calcium, vitamin C, iron, and sodium in the one-day menu that you constructed.

5. The National Cholesterol Education Program (NCEP) has recommended the following guidelines as a first step toward lowering blood-cholesterol levels:

Cholesterol	< 300 mg
Saturated fat	< 10 percent of total calories
Fat	< 30 percent of total calories

Construct a spreadsheet, or add to one already constructed, to calculate the percent of calories from fat and from saturated fat and the amount of cholesterol in a given menu. The table on the next page, adapted from the *Consumer Guide FAT Counter Guide,* 1992, gives some examples. Add other food items if desired.

6. Use your spreadsheet to calculate the percent of calories from fat and from saturated fat and the amount of cholesterol in the sample menu given at the beginning of this activity.

7. Use your spreadsheet to calculate the percent of calories from fat and from saturated fat and the amount of cholesterol in the sample menu you constructed in task 2.

8. By experimenting with amounts and types of food in your menu, present an optimal and interesting one-day menu that (1) contains as much food from each food category as is recommended by the *New American Eating Guide,* (2) corresponds to the recommended dietary allowance for someone of your age and gender (obtain chart from your teacher), and (3) follows the NCEP dietary recommendations for lowering blood cholesterol.

9. Write a report on your experimentation. Include in your report—

   • the first menu you constructed;
   • a description of how you changed your menu to fit the national recommendations;
   • a description of any insights you had or patterns you detected regarding the effects of your changes;
   • any advice you would give someone who wanted to construct a menu corresponding to both sets of guidelines.

Item	Amount	Total Fat (grams)	Fat (as % of calories)	Total Saturated Fat (grams)	Saturated Fat (as % of calories)	Cholesterol (milligrams)
Banana	1	1	9	<1	2	0
Bread (rye)	1 slice	1	14	<1	3	0
Bread (whole wheat)	1 slice	1	13	n/a	n/a	0
Broccoli	1 cup	trace	trace	<1	2	0
Butter	1 tsp	5	99	3	63	15
Carrots	1 cup	trace	n/a	trace	n/a	0
Cheese (Swiss)	1 slice	7	70	4	40	25
Chocolate chips	1/6 cup	8	49	5	29	5
Egg (scrambled)	1	7	63	2	18	215
Ham (cooked)	0.25 lb	7	45	2	13	35
Hamburger	0.25 lb	18	66	7	26	76
Lettuce	1 cup	trace	n/a	trace	n/a	0
Mayonnaise	1 Tbsp	11	99	2	18	7
Milk (low fat)	1 cup	5	26	3	22	18
Milk (skim)	1 cup	1	9	trace	<1	5
Mustard	1 Tbsp	1	82	0	0	0
Peanuts (salted)	1/6 cup	11	66	2	9	0
Pork chop (baked)	0.25 lb	12	50	5	20	88
Potato (baked)	1	trace	n/a	trace	n/a	0
Raisins	1 cup	1	12	<1	<1	0
Raspberries	1 cup	1	15	trace	n/a	0
Roll (hard)	1	2	12	<1	2	trace
Steak (T-bone)	0.25 lb	6	36	3	18	64

ACTIVITY 13

REFRESHMENT-STAND PROFITS

The Central High School band raises money each year to defray the cost of its annual trip to Disney World. Its major fund-raising activity is running the refreshment stands on weekends at Funworld, the regional amusement park. Local businesses contribute all the refreshments and other supplies, but the band must pay $110 a week to rent the stands for the twelve-week season.

In the course of analyzing the prospects for making money, band members observed that—

- the number of admission tickets sold seemed to be a function of the price of an admission ticket;
- the band's profit seemed to be a function of the number of admission tickets sold;
- on the average, refreshment-stand revenue seemed to be approximately $4.75 for each paid admission.

To help them analyze the situation, they conducted a survey of potential customers. The data they collected are shown below. In this example, the row "InData = 4, OutData = 2703" means that 2703 people would buy admission tickets if the price of a ticket were $4.00.

InData	OutData
2	3375
4	2703
6	2095
8	1430
10	817

1. The number of admission tickets sold seems to be a function N of the price of an admission ticket t. Use a curve-fitting program to find a rule that describes the relationship between the number of tickets sold and the price of a ticket. Record your function rule here:

   $N(t) =$ ______________________________

   Sketch a graph of the survey data along with your best-fitting function rule. Label the axes.

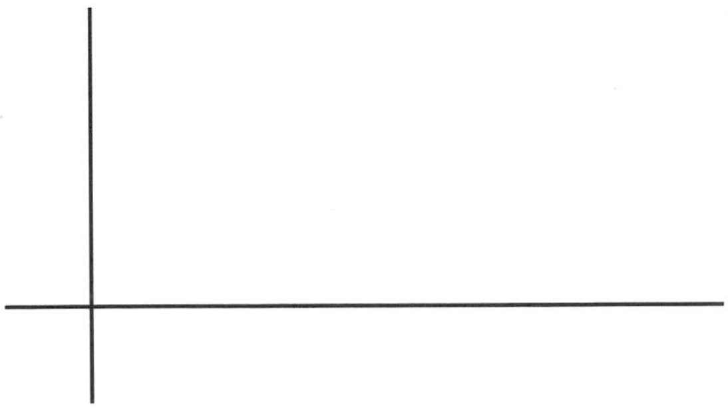

2. The profit for the twelve-week season is a function of the number of admission tickets sold. If *a* represents the number of tickets sold, write a function rule for profit P_1 as a function of *a*.

    $P_1(a) =$ _____

    Use your best-fitting function rule for $N(t)$ to determine the number of tickets sold for different prices, then use the numbers of tickets sold to determine profit. Record those results in the following table.

Price of a ticket ($)	Number of people who would buy a ticket at this price	Band's profit if this many tickets are sold
3.00		
4.50		
6.00		
7.50		
9.00		
10.50		
12.00		

3. Instead of first calculating the number of tickets sold at a given price, it is possible to calculate the profit directly from the price of a ticket. Figure out how to do this calculation and be ready to explain your procedure and your reasoning. (Hint: Use the computer or your knowledge of symbolic rules to "simplify" Profit = $P(N(t))$.)

4. What is the best ticket price if the Central High School band is to make as much money as possible? Explain your reasoning.

ACTIVITY 14

EXPLORING FUNCTION ORBITS

Explore orbits for each of the following functions by using a calculator, a computer, or a spreadsheet.

1. Let $f(x) = x^2$. What happens to the orbits if—

   (a) $|x_0| > 1$?

   (b) $|x_0| < 1$?

   (c) $|x_0| = 1$?

2. Let $g(x) = \cos(x)$. What happens to the orbits for any x_0?

3. Let $h(x) = \dfrac{x + \dfrac{4}{x}}{2}$.

   (a) What happens to the orbits for $x_0 > 0$? For $x_0 < 0$?

   (b) Examine the orbits for

   $$k(x) = \frac{x + \dfrac{9}{x}}{2}$$

   and

   $$m(x) = \frac{x + \dfrac{2}{x}}{2}.$$

   Do you notice any patterns?

CHAPTER 5
MATRICES

Most calculators and computers have the capability to do matrix algebra. Matrices are invaluable mathematical tools that have a variety of uses. Like real numbers, they can describe quantitative phenomena and have an arithmetic and an algebra that can be applied to solve problems. Like numbers, matrices have an interesting algebraic structure that tells a lot about how they behave. Like functions on real numbers, the functions on matrices are useful for modeling real-life situations.

Matrices have been around for a long time, but only recently, with the arrival of inexpensive, easy-to-use computers and graphing calculators, have they become accessible to the beginning high school student. Presented below are some matrix algebra topics and applications that are appropriate for the high school classroom in a technological world.

MATRIX ALGEBRA

Planning Airplane Routes

In planning an airline trip, it is often impossible to book a flight directly from one city to another. For example, suppose we want to fly from Portland, Maine, to Baltimore, Maryland. We would have to make the trip via another city, say Boston, because no direct Portland-to-Baltimore flights are available. Planning a trip can become quite complicated, but matrices make the process easier. Consider a hypothetical airline, Air New England, which uses terminals only in the following cities: Burlington, Vermont; Hartford, Connecticut; Worcester, Massachusetts; Providence, Rhode Island; and Manchester, New Hampshire. The airline offers the following flights:

Burlington to Hartford, Manchester, and Worcester

Hartford to Manchester and Portland

Manchester to Burlington, Hartford, Portland, and Worcester

Portland to Manchester and Worcester

Worcester to Burlington and Portland

This list gives all the flights but not in a very accessible form. A better method is to put the information in graphical form, such as the diagram shown in figure 5.1. This presentation tells us more but still makes it difficult to get much information about all possible flights.

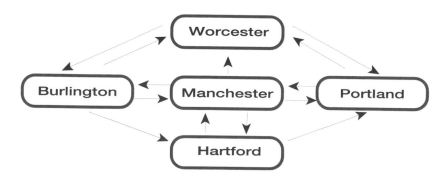

Fig. 5.1. An Air New England route diagram

A possibly better way is to put the information into an array or matrix where the row name indicates the departure city and the column name indicates the arrival city. This matrix is given in figure 5.2. Note that a 1 is the entry for the cell in the row headed Manchester and the column headed Hartford because there is a flight from Manchester to Hartford. The entry in the cell that has its row labeled Burlington and its column labeled Portland is O because there is no direct flight from Burlington to Portland. This matrix representation is an improvement on the pictorial representation in figure 5.1 because the matrix can be used more easily to generate new information about the airline routes.

Arrival City

		Burlington	Hartford	Manchester	Portland	Worcester
	Burlington	O	1	1	O	1
	Hartford	O	O	1	1	O
Departure City	Manchester	1	1	O	1	1
	Portland	O	O	1	O	1
	Worcester	1	O	O	1	O

Fig. 5.2. A matrix giving cities served by Air New England

Classes examining this matrix might be asked questions like these:

♦ How many nonzero entries are there? What does the total number of nonzero entries represent?

♦ What does the total number of 1s in any one column represent?

♦ What does the total number of 1s in any one row represent?

♦ Why is the matrix not symmetric about the main diagonal?

♦ Why are the entries along the main diagonal all O's?

An interesting question is this: In how many ways can we fly from Burlington to Portland? Looking at the matrix, we can see that there is no way to get there directly. However, we could get there by first flying from Burlington to an intermediate city and then on to Portland. Since flights are possible from Burlington to Hartford, Manchester, and Worcester and from each of these cities to Portland, there are three flight routes from Burlington to Portland with one stop. To figure out all the one-stop routes would be tedious to do by hand but can easily be done using matrix multiplication. To calculate the total number of one-stop B to P routes, we need to find the sum of the following products. Have (X to Y) denote the number of routes from city X to city Y.

(B to B)(B to P) + (B to H)(H to P) + (B to M)(M to P) + (B to P)(P to P)

+ (B to W)(W to P)

= (O)(O) + (1)(1) + (1)(1) + (O)(O) + (1)(1)

= 3

But this calculation is just the inner product of the row labeled B and the column headed P and hence is the entry in the Bth row and the Pth column of the product. Therefore, to calculate the number of one-stop trips among the five cities, all we need do is multiply the flight-information matrix by itself. Doing this, we get the result shown in figure 5.3.

	B	H	M	P	W			B	H	M	P	W			B	H	M	P	W
B	0	1	1	0	1		B	0	1	1	0	1		B	2	1	1	3	1
H	0	0	1	1	0		H	0	0	1	1	0		H	1	1	1	1	2
M	1	1	0	1	1	×	M	1	1	0	1	1	=	M	1	1	3	2	2
P	0	0	1	0	1		P	0	0	1	0	1		P	2	1	0	2	1
W	1	0	0	1	0		W	1	0	0	1	0		W	0	1	2	0	2

Fig. 5.3. The matrix product that gives the number of one-stop flight routes

This product tells us, among other things, that one one-stop route from Portland to Worcester and no one-stop routes from Worcester to Burlington or from Portland to Manchester are available to us.

The following questions can be answered using matrix methods similar to the one in figure 5.3.

◆ How many one-stop flight routes are there from Portland to Burlington? From Manchester to Portland? From Manchester to Manchester?

◆ Find the number of two-stop flight routes possible on Air New England.

◆ Discuss the applicability of this method to an actual flight schedule. In what ways does the matrix model of the situation fit the real world? In what ways does this model not fit the real world?

For another way to encourage the study of matrices and matrix operations, see *Connecting Mathematics* (Froelich et al. 1991), also from the Grades 9–12 Addenda series.

Keeping Track of Rental Cars

Matrix algebra can be used to study the behavior over time of systems that consist of several parts. For example, consider the Hello Good-Bye Rental Car Company. They have locations in three cities: A, B, and C. Their cars will often start in one city and end up in another. The company wants to have some way of predicting how many cars will be in each location at the end of each month. Their research on where cars start each month and where they end up is summarized in matrix T (see fig. 5.4). The entry in the ith row and the jth column gives the percent of cars that start in city j and end up in city i. In the following example, the entry in row B and column A is 0.2, which means that 20 percent of the cars that start in A end up in B.

$$T = \text{End } \begin{matrix} A \\ B \\ C \end{matrix} \begin{bmatrix} 0.35 & 0.1 & 0.0 \\ 0.20 & 0.2 & 0.8 \\ 0.45 & 0.7 & 0.2 \end{bmatrix}$$

Start: A B C

Fig. 5.4. Transition matrix T for the Hello Good-Bye Rental Car Company

This matrix T, called a *transition matrix*, describes how the percent of cars in the different cities changes over time. For example, suppose that at the beginning of the year, 200 cars were in location A, 500 in location B, and 300 in location C. The matrix X_0 described below contains this information. To find out how many cars were in each location at the end of the first month, X_1, we form the product TX_0. To find X_2, the number of cars in each location at the end of two months, calculate TX_1. Notice that $X_2 = T(X_1) = T(TX_0)$. Since

$$T = \begin{bmatrix} 0.35 & 0.10 & 0 \\ 0.20 & 0.20 & 0.80 \\ 0.45 & 0.70 & 0.20 \end{bmatrix} \quad \text{and} \quad X_0 = \begin{bmatrix} 200 \\ 500 \\ 300 \end{bmatrix},$$

$$X_1 = TX_0 = \begin{bmatrix} 0.35 & 0.10 & 0 \\ 0.20 & 0.20 & 0.80 \\ 0.45 & 0.70 & 0.20 \end{bmatrix} \begin{bmatrix} 200 \\ 500 \\ 300 \end{bmatrix} = \begin{bmatrix} 120 \\ 380 \\ 500 \end{bmatrix}.$$

We can calculate X_2 similarly:

$$X_2 = TX_1 = \begin{bmatrix} 0.35 & 0.10 & 0 \\ 0.20 & 0.20 & 0.80 \\ 0.45 & 0.70 & 0.20 \end{bmatrix} \begin{bmatrix} 120 \\ 380 \\ 500 \end{bmatrix} = \begin{bmatrix} 80 \\ 500 \\ 420 \end{bmatrix}$$

This calculation shows that after two months, 80 cars will be in location A, 500 in location B, and 420 in location C. In a similar fashion, we can find the numbers after three months, after four months, and so on. Using a computer or a matrix-capable calculator makes this question very easy to answer. Shown in the margin is an algorithm for the TI-81 or TI-82 graphing calculator that can keep track of the rental cars.

Here are some questions that can be answered using matrices generated on a calculator:

♦ How many cars will be in each location after five months?

♦ Suppose that at the outset, 100 cars were at location A, 100 at location B, and 800 at location C. How many cars would be at each location after 1 month, 2 months, 4 months, 6 months, and 1 year?

♦ Is it possible to find numbers a, b, and c such that if a cars were at location A, b cars at location B, and c cars at location C, then at the end of the month the same number would be at each location?

Further ideas on how matrices can be used to model qualitative as well as quantitative aspects of real-world situations can be found in *A Core Curriculum: Making Mathematics Count for Everyone* (Meiring et al. 1992), a volume in the Grades 9–12 Addenda series.

SOLVING LINEAR SYSTEMS

There are times when it is helpful to solve systems of linear equations. The following situation, from *Concepts in Algebra: A Technological Approach* (Fey and Heid 1995, p. 360), provides an example.

Rental-Car Algorithm

♦ **Enter the matrix T as matrix A and the matrix X_0 as B and then enter QUIT.**

♦ **Enter [A][B], and press ENTER. This will give the 3 ¥ 1 matrix X1, which gives the number of cars at each location at the end of the first month.**

♦ **Enter [A] ANS, and press ENTER. This will give the 3 × 1 matrix X2, which gives the number of cars at each location at the end of the second month.**

♦ **Keep pressing ENTER to get a new array giving the number of cars at each location.**

◆　　◆　　◆　　◆　　◆　　◆　　◆　　◆

> ### Baseball-Game Promotion
>
> A baseball team is planning a special promotion at one of its games. A fan who arrives early will get a team athletic bag or a jacket, as long as the supply lasts. The promotion manager knows that each athletic bag costs $5.50 and each jacket costs $9.75. What combination of team bags and jackets can provide for a total of 3 500 fans at a cost of $23 500.00?

To answer the question posed about the baseball promotion, the following system of linear equations needs to be solved, where b represents the number of bags and j represents the number of jackets:

$$5.50b + 9.75j = 23\ 500$$
$$b + j = 3\ 500$$

Matrices offer a method for solving systems of linear equations like these. Here is a way to solve this system using matrices generated by a calculator. First, express the system in matrix form. Let

$$A = \begin{bmatrix} 5.50 & 9.75 \\ 1 & 1 \end{bmatrix}, \quad X = \begin{bmatrix} b \\ j \end{bmatrix}, \quad \text{and } B = \begin{bmatrix} 23\ 500 \\ 3\ 500 \end{bmatrix}.$$

Then

$$5.50b + 9.75j = 23\ 500$$
$$b + j = 3\ 500$$

is equivalent to $AX = B$.

As long as A, which is called the *coefficient matrix,* has an inverse, we can multiply both sides of $AX = B$ by A^{-1} to obtain an equivalent equation.

$$AX = B$$
$$\Rightarrow \quad A^{-1}AX = A^{-1}B$$
$$\Rightarrow \quad IX = A^{-1}B$$
$$\Rightarrow \quad X = A^{-1}B$$

Now using a calculator, enter matrices A and B and then calculate $A^{-1}B$. Performing these steps for the baseball-promotion system of equations, we get an output that corresponds to

$$X = \begin{bmatrix} 2\ 500 \\ 1\ 000 \end{bmatrix},$$

suggesting a feasible solution of 2 500 team bags and 1 000 jackets.

When starting to analyze a situation like this, we are unlikely to know exactly the total amount of money we will spend. Instead, we are more likely to know our spending limit. For example, in this situation, we are more likely to know that we want to plan for 3 500 fans and that we can spend no more than $25 000 than know that we want to spend exactly $23 500 for 3 500 fans. We can use the solution of a related system of linear equations as a starting point. The related system of equations is

$$5.50b + 9.75j = 25\ 000$$
$$b + j = 3\ 500,$$

with corresponding matrices

$$A = \begin{bmatrix} 5.50 & 9.75 \\ 1 & 1 \end{bmatrix} \quad \text{and } B = \begin{bmatrix} 25\ 000 \\ 3\ 500 \end{bmatrix}.$$

Then $AX = B$ and

$$X = A^{-1}B = \begin{bmatrix} 5.50 & 9.75 \\ 1 & 1 \end{bmatrix}^{-1} \begin{bmatrix} 25\ 000 \\ 3\ 500 \end{bmatrix} = \begin{bmatrix} 2\ 147.058\ 824\ 1 \\ 1\ 352.941\ 176\ 1 \end{bmatrix}.$$

This calculation suggests 2 147 bags and 1 353 jackets. We can explore this answer in more depth by examining nearby values.

Using a spreadsheet or a tables program like that for a graphing calculator shown in the margin, we can calculate the costs of various options. We find that purchasing 2 147 bags and 1 353 jackets would cost $25 000.25. We need to spend a little less, so we will increase the number of bags, the less expensive item, while decreasing the number of jackets. Purchasing 2 148 bags and 1 352 jackets costs $24 996.00. If we had decreased the number of bags to 2 146 instead, the cost of 2 146 bags and 1 354 jackets would have increased to $25 004.50. These results are displayed in the chart below:

Number of Bags	Number of Jackets	Cost
2 148	1 352	$24 996.00
2 147	1 353	$25 000.25
2 146	1 354	$25 004.50

Notice that for every additional jacket purchased, one fewer bag is purchased and the cost increases by $4.25, the difference in cost between a bag and a jacket.

Tables Program for a Graphing Calculator

Prgm 1: PROMO

:Lbl 1

:Disp "BAGS"

:Input B

:Disp "JACKETS"

:3500 – B –> J

:Disp J

*:5.5 * B + 9.75 * J –> C*

:Disp C

:Goto 1

Assessment Matters: A good way to assess if students understand the meaning of the entries in a matrix is to give them a particular matrix and describe several real-world situations that the matrix could model, such as planning airplane routes, keeping track of rental cars, and solving systems of linear equations. Ask them to explain the meaning of each entry in the matrix.

Connections between Matrices and the Geometry of Systems of Linear Equations

Some interesting mathematical connections exist between geometric and matrix representations of systems of equations. We will look at some of those connections by examining the conditions under which a system has a unique solution.

From a matrix point of view, the linear system $AX = B$ has a unique solution if and only if matrix A has an inverse. That is, the linear system

$$ax + by = e$$
$$cx + dy = f$$

has a unique solution if and only if the coefficient matrix $A = \begin{bmatrix} a & b \\ c & d \end{bmatrix}$ has an inverse.

From a geometric point of view, when we solve a 2×2 system of linear equations, we are finding the point of intersection of the graphs of two lines. If the two lines intersect, the system of equations has a unique solution; if the lines are parallel, the system has no solution; and if the lines coincide, the system has infinitely many solutions. If we are solving the system

$$ax + by = e$$
$$cx + dy = f,$$

then we are finding the intersection of the two lines

$$y = \frac{-a}{b}x + \frac{e}{b} \quad \text{and} \quad y = \frac{-c}{d}x + \frac{f}{d}.$$

These lines are parallel or coincide if and only if their slopes are equal. That is,

$$\frac{-a}{b} = \frac{-c}{d},$$

or, equivalently, $ad - bc = 0$.

Putting these two points of view together, we see that the matrix
$\begin{bmatrix} a & b \\ c & d \end{bmatrix}$ has an inverse if and only if $ad - bc \neq 0$. So the system

$$ax + by = e$$
$$cx + dy = f$$

has a unique solution if and only if $ad - bc \neq 0$. The quantity $ad - bc$ is important enough to have a name. It is called the *determinant* of matrix A and is denoted det[A]. So the system $AX = B$ will have a unique solution if and only if det[A] $\neq 0$.

If det[A] = 0, the system has two possibilities: no solutions or infinitely many solutions. We can use matrices to distinguish between these two cases. Take the system

$$ax + by = e$$
$$cx + dy = f$$

and solve each equation for y, getting

$$y = \frac{-a}{b}x + \frac{e}{b} \quad \text{and} \quad y = \frac{-c}{d}x + \frac{f}{d}.$$

If the lines coincide,

$$\frac{-a}{b} = \frac{-c}{d} \quad \text{and} \quad \frac{e}{b} = \frac{f}{d},$$

or, equivalently, $ad - bc = 0$ and $ed - fb = 0$. If the lines do not coincide but are parallel,

$$\frac{-a}{b} = \frac{-c}{d} \quad \text{and} \quad \frac{e}{b} \neq \frac{f}{d},$$

or, equivalently, $ad - bc = 0$ and $ed - fb \neq 0$. We can think of $ed - fb$ as the determinant of the matrix

$$A' = \begin{bmatrix} e & b \\ f & d \end{bmatrix},$$

which is the matrix formed by replacing the first column of the coefficient matrix A with the constant matrix B.

We now have a more complete picture of solving systems of linear equations by using matrices. In summary, to solve the system

$$ax + by = e$$
$$cx + dy = f,$$

follow these steps:

1. Form the three matrices: A (the coefficient matrix); B (the constant matrix); and A' (the matrix formed from A by replacing the first column of A with the only column in B).

2. Consult the chart in figure 5.5, which indicates possible solutions.

Assessment Matters: Give students a copy of such a flowchart when they are being assessed. If the teacher is interested primarily in their ability to solve modeling problems, it may not be important for them to reproduce this flowchart.

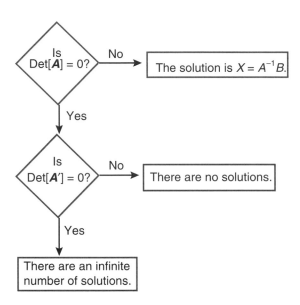

Is Det[A] = 0? No → The solution is $X = A^{-1}B$.

Yes ↓

Is Det[A'] = 0? No → There are no solutions.

Yes ↓

There are an infinite number of solutions.

Fig. 5.5. Flowchart for solving a 2 × 2 system of linear equations

GEOMETRIC FUNCTIONS

For most of the functions discussed in previous chapters, both the input and the output have been numbers. We can, as well, have functions that have matrices both as inputs and as outputs. Studying these functions with the aid of technology can lead to some important connections between algebra and geometry.

Function with Rule F[X] = X + B

For example, suppose we had a function F that took points and moved each of them four units to the right. For this function, as shown in figure 5.6,

(–2, 3)	maps to	(2, 3);
(–2, 7)	maps to	(2, 7); and
(1, 3)	maps to	(5, 3).

This function can be described by the rule

$F[X] = X + B$, where X is a 2 × 1 input matrix $\begin{bmatrix} x \\ y \end{bmatrix}$ and B = $\begin{bmatrix} 4 \\ 0 \end{bmatrix}$.

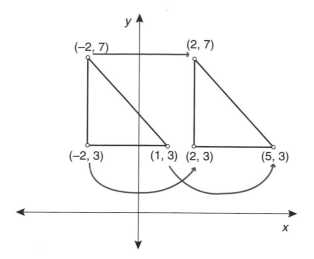

Fig. 5.6. Illustration of a function F that takes points and moves each of them four units to the right

This function has a 2 × 1 matrix as both input and output. As a second example, let G be the function that transforms any point X into the point X + B, where

$$B = \begin{bmatrix} 2 \\ 3 \end{bmatrix}.$$

In this example, if

$$X = \begin{bmatrix} 1 \\ 1 \end{bmatrix},$$

then

$$X + B = \begin{bmatrix} 3 \\ 4 \end{bmatrix}.$$

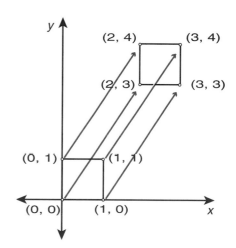

Fig. 5.7. Unit square translated by

$$B = \begin{bmatrix} 2 \\ 3 \end{bmatrix}$$

The function G has the effect of translating X by two units horizontally and three units vertically. To study functions of this form, it is worthwhile seeing what the function does to a set of points such as those in a square. In the following example, the set S will refer to the square with vertices at A(0, 0), B(1, 0), C(1, 1), and D(0, 1). We will denote the image of S by G[S] = S'. In this example, G translates S by B (see fig. 5.7). In general, any function of the form G[X] = X + B is a translation.

Function with Rule F[X] = KX

Next consider functions of the form F[X] = KX, where K is a 2 × 2 matrix. Here the story is far more complicated. In this section is a sampling of the different effects that F can have on the unit square S as different matrices K are used. A graphing calculator makes it easy to compute the image of the unit square (A(0, 0), B(1, 0), C(1, 1), and D(0, 1)). One way is to calculate the output of F for each of the four vertices, A, B, C, and D. We can shortcut the process, however, by forming a 2 × 4 matrix.

First, look at the effects of a seemingly arbitrarily chosen matrix K on the matrix formed by the vertices of the unit square.

Let $K = \begin{bmatrix} 1 & 1 \\ -1 & 1 \end{bmatrix}$. Let $X = \begin{bmatrix} 0 & 1 & 1 & 0 \\ 0 & 0 & 1 & 1 \end{bmatrix}$, where the columns of X are the vertices of the unit square S. Compute $S' = KX$.

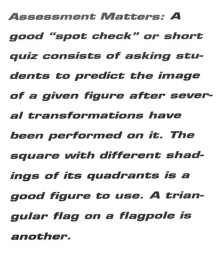

Then,

$$KX = \begin{bmatrix} 1 & 1 \\ -1 & 1 \end{bmatrix}\begin{bmatrix} 0 & 1 & 1 & 0 \\ 0 & 0 & 1 & 1 \end{bmatrix} = \begin{bmatrix} 0 & 1 & 2 & 1 \\ 0 & -1 & 0 & 1 \end{bmatrix},$$

which gives the coordinates of the image of the unit square ([0, 0], [1, −1], [2, 0], and [1, 1]). The unit square and its image are plotted in figure 5.8.

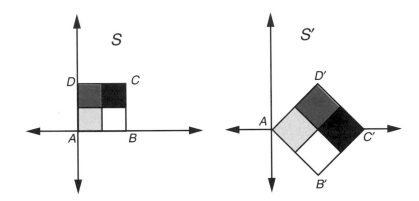

Fig. 5.8. The unit square S and its image under transformation by F[X] = KX

Comparing S' to S, we see that K has dilated S by a factor of $\sqrt{2}$ then rotated it 45 degrees clockwise around A. If we were to multiply S' by K, the result would be square $AB''C''D''$, which was formed by dilating S' by a factor of $\sqrt{2}$, then rotating it 45 degrees clockwise around A (see fig. 5.9).

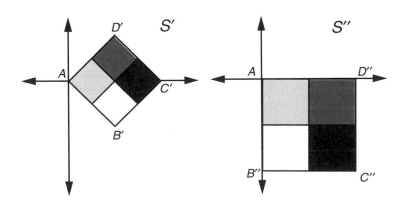

Fig. 5.9. S″ is the image of S′ under the transformation F[X] = KX

We can also go the other way and ask what set of points L are the input for an output of S. To answer, we must solve the matrix equation

$$KL = S,$$

where S is the vertex arrangement of the original square. The solution is $L = K^{-1}S$ as long as K has an inverse. Using the matrix commands on a graphing calculator, we can easily find the solution:

$$L = K^{-1}S,$$

or

$$L = \begin{bmatrix} 1 & 1 \\ -1 & 1 \end{bmatrix}^{-1} \begin{bmatrix} 0 & 1 & 1 & 0 \\ 0 & 0 & 1 & 1 \end{bmatrix} = \begin{bmatrix} 0 & 0.5 & 0 & -0.5 \\ 0 & 0.5 & 1 & 0.5 \end{bmatrix}.$$

Since K^{-1} is the inverse of K, it must undo whatever K does. Hence, K^{-1} rotates S counterclockwise 45 degrees around A and then shrinks the resulting figure by a factor of $1/\sqrt{2}$.

Technology makes it easy to plot the following sequence of shapes,

$$S, KS, K^2S, K^3S, K^4S, \ldots,$$

which form a *clockwise* spiral of *expanding* and rotating squares as shown in figure 5.10.

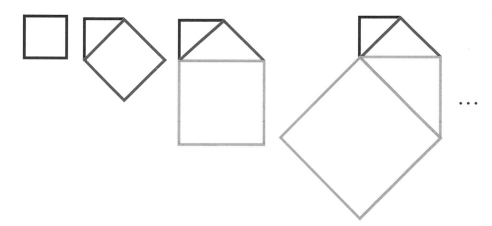

Fig. 5.10. A clockwise spiral of expanding and rotating squares

Similarly, the sequence of shapes

$$S, K^{-1}S, K^{-2}S, K^{-3}S, K^{-4}S, \ldots$$

form a *counterclockwise* spiral of *shrinking* and rotating squares as shown in figure 5.11.

Fig. 5.11. A counterclockwise spiral of shrinking and rotating squares

Teaching Matters: The con-sideration of the reflection over the line y = 2x was developed through a combi-nation of by-hand symbolic manipulation and comput-ing-tool use. It is important to note that the by-hand symbolic manipulation drew only on very basic under-standings of the meaning of the mathematical concepts and procedures. This is one model of the integration of by-hand work with comput-ing technology. As algebra enters the twenty-first cen-tury, the mathematics-education community must come to terms with appro-priate roles for by-hand symbolic manipulation.

The final example in this section illustrates the interplay between the conceptual understanding of ideas of algebra and geometry and the thoughtful use of computing technology.

Reflection over the Line $y = 2x$

If possible, find a 2×2 matrix K so that $F[X] = KX$ reflects a figure over the line $y = 2x$.

The matrix for this function can be derived using two properties of reflec-tions. Both properties are derived from the fact that if P maps to P' in reflection, the line of reflection is the perpendicular bisector of $\overline{PP'}$.

Let

$$P = \begin{vmatrix} x \\ y \end{vmatrix}$$

and

$$P' = \begin{bmatrix} x' \\ y' \end{bmatrix}.$$

Since $\overline{PP'}$ must be perpendicular to $y = 2x$, then the slope of $\overline{PP'}$ is the negative reciprocal of the slope of $y = 2x$ and, therefore,

$$\frac{y - y'}{x - x'} = -\frac{1}{2}.$$

So,

$$2(y - y') = -(x - x')$$
$$2y - 2y' = x' - x$$
$$x' + 2y' = x + 2y.$$

Since the midpoint

$$\begin{bmatrix} (x + x')/2 \\ (y + y')/2 \end{bmatrix}$$

Try This: Find a 2 × 2 matrix K such that F[X] = KX reflects a figure over the y-axis.

of PP' must lie on the line $y = 2x$, it must be true that the y-value of the midpoint is twice its x-value. Hence,

$$\frac{y + y'}{2} = 2\left(\frac{x + x'}{2}\right)$$
$$y + y' = 2x + 2x'$$
$$2x' - y' = y - 2x.$$

Try This: Find a 2 × 2 matrix K such that F[X] = KX rotates a figure 180 degrees about the origin.

We can find appropriate values for x' and y' if we can solve the system

$$x' + 2y' = x + 2y$$
$$2x' - y' = y - 2x.$$

In matrix notation, this system can be written

$$\begin{bmatrix} 1 & 2 \\ 2 & -1 \end{bmatrix}\begin{bmatrix} x' \\ y' \end{bmatrix} = \begin{bmatrix} 1 & 2 \\ -2 & 1 \end{bmatrix}\begin{bmatrix} x \\ y \end{bmatrix}.$$

$$KP' = BP$$

$$P' = K^{-1}BP$$

$$P' = \begin{bmatrix} 1 & 2 \\ 2 & -1 \end{bmatrix}^{-1}\begin{bmatrix} 1 & 2 \\ -2 & 1 \end{bmatrix}P$$

$$P' = \begin{bmatrix} -0.6 & 0.8 \\ 0.8 & 0.6 \end{bmatrix}P.$$

Therefore,

$$F[X] = \begin{bmatrix} -0.6 & 0.8 \\ 0.8 & 0.6 \end{bmatrix}X.$$

Function with Rule F[X] = KX + B

At times, we want to apply transformations that are not defined with respect to the origin. For example, suppose we wanted to rotate a point $X(x, y)$ counterclockwise through 90 degrees about the point $P(2, 4)$.

Geometrically, we could translate the point through the directed distance $\overrightarrow{PO}$, rotate the result counterclockwise through 90 degrees, and translate back by $\overrightarrow{OP}$. Figure 5.12 shows this sequence of transformations applied to the point $(-3, 1)$.

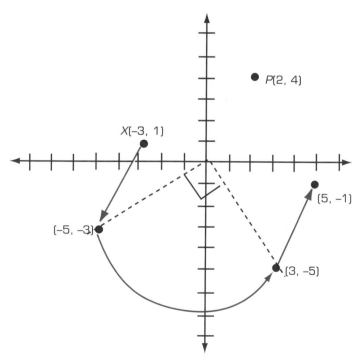

Fig. 5.12. Rotation of a point X counterclockwise about the point P(2, 4) through 90 degrees can be accomplished by first translating X through the directed distance $\overrightarrow{PO}$, rotating the result 90 degrees counterclockwise about the origin, and then translating this result by the vector $\overrightarrow{OP}$.

Using matrices, we have seen that $F[X] = X + B$ translates points by B. We could also show that

$$F[X] = \begin{bmatrix} 0 & -1 \\ 1 & 0 \end{bmatrix} X$$

rotates a figure counterclockwise through 90 degrees. So the desired function is

$$F[X] = \begin{bmatrix} 0 & -1 \\ 1 & 0 \end{bmatrix} \left(X - \begin{bmatrix} 2 \\ 4 \end{bmatrix} \right) + \begin{bmatrix} 2 \\ 4 \end{bmatrix}.$$

Note that F does nothing to

$$\begin{bmatrix} 2 \\ 4 \end{bmatrix}$$

and hence is called a *fixed point*. Any function of the form $F[X] = K(X - P) + P$ is said to be in *fixed-point form*.

Reexpressing $F[X] = K(X - P) + P$, we get $KX - KP + P = KX + B$, where $B = P - KP$. This shows that rotations about points other than the origin can always be expressed in the form $F[X] = KX + B$.

Conversely, we can show that a function of the form $KX + B$ has the same geometric effect as KX with respect to the point P instead of the origin. This means that to understand what the function $F[X] = KX + B$ does, we really need only know what K does and what point is fixed.

Further work with matrix representations of transformations can be found in *Geometry from Multiple Perspectives* (Coxford et al. 1991), another book in the Grades 9–12 Addenda series.

CONCLUSION

Because of the ready availability of technology, matrices can now be easily used by beginning high school students. The use of matrices as organizational tools, as modeling tools, as geometric functions, and as tools for extending graphing capabilities should enrich the algebra strand of the curriculum in years to come. Their use in geometry should provide an important bridge between these two very important fields of mathematics.

ACTIVITY 15

GEOMETRIC FUNCTIONS FOR SPECIFIC TASKS

1. The function $F[X] = KX$, where

$$K = \begin{bmatrix} 1 & 1 \\ -1 & 1 \end{bmatrix},$$

   when applied to the matrix for the unit square with vertices $A(0, 0)$, $B(1, 0)$, $C(1, 1)$, and $D(0, 1)$, dilates the unit square by a factor of $\sqrt{2}$, then rotates it 45 degrees clockwise around the origin. Verify this statement.

2. Use the matrix operations on your calculator to determine what happens to the unit square when it is multiplied by each of the following 2×2 matrices.

$$A = \begin{bmatrix} 1 & 0 \\ 0 & -1 \end{bmatrix} \quad B = \begin{bmatrix} -1 & 0 \\ 0 & -1 \end{bmatrix} \quad C = \begin{bmatrix} -1 & 0 \\ 0 & 1 \end{bmatrix} \quad D = \begin{bmatrix} 0 & 1 \\ 1 & 0 \end{bmatrix}$$

$$E = \begin{bmatrix} 1 & -1 \\ -1 & 0 \end{bmatrix} \quad F = \begin{bmatrix} 1 & -1 \\ 1 & 1 \end{bmatrix} \quad G = \begin{bmatrix} 0 & 1 \\ -1 & 0 \end{bmatrix} \quad H = \begin{bmatrix} 0 & -1 \\ 1 & 0 \end{bmatrix}$$

   Each 2×2 matrix performs its own unique transformation on the unit square.

3. A matrix transformation that preserves distances is called an *isometry*. Which of the preceding matrix transformations seem to be isometries?

4. When a point matrix is multiplied by a 2×2 matrix, some points stay where they are and some move. Identify the fixed points for each matrix in task 2.

5. Investigate what happens to the unit square when its matrix is multiplied by each of the following products of matrices:

   (a) AB

   (b) BC

   (c) DF

   (d) EF

   (e) FE

WHAT DOES F[S] = KS DO TO THE AREA OF THE UNIT SQUARE S?

1. If $K = \begin{bmatrix} 1 & 2 \\ 3 & -2 \end{bmatrix}$, $KS = \begin{bmatrix} 1 & 2 \\ 3 & -2 \end{bmatrix}\begin{bmatrix} 0 & 1 & 1 & 0 \\ 0 & 0 & 1 & 1 \end{bmatrix} = \begin{bmatrix} 0 & 1 & 3 & 2 \\ 0 & 3 & 1 & -2 \end{bmatrix}$.

The effect of multiplying the unit square by K is shown in the diagram below.

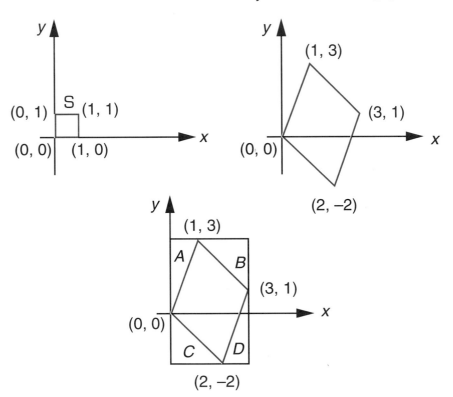

The unit square

The unit square transformed by the function $F[S] = KS$

Area of S' = Area of rectangle – Areas of the four right triangles

(a) Find the area of the "enclosing" rectangle.

(b) Find the area of each right triangle.

(c) Find the area of quadrilateral S'.

(d) How is the area of S' related to the value of the determinant of K? Test your conjecture on a few other transformation matrices.

(e) Choose a different transformation matrix K and find the area of the image of the unit square S under this transformation. How is the area of S' related to the value of the determinant of K?

2. What is the effect of any transformation matrix K on the area of R', where R is a rectangle different from the unit square?

3. What is the effect of any transformation matrix K on the area of P', where P is a nonrectangular parallelogram?

4. Make and test a conjecture about the effect on the area of the image of Q of $F[Q] = KQ$ when K is any 2×2 matrix and Q is any quadrilateral.

CHAPTER 6
SYMBOLIC REASONING

In a technological world, symbolic manipulation retains many of its traditional roles. The difference is that the symbolic manipulation of the future will not generally be done by hand.

SYMBOLIC MANIPULATION IN TRADITIONAL ALGEBRA CLASSES

Algebra courses in schools have historically centered on the by-hand production of equivalent forms of expressions and equations. In particular, students learned the rules needed to produce equivalent forms for polynomial, rational, and exponential expressions. For example, they learned procedures that enabled them to determine that for real numbers x—

♦ $2x^2 + 5x - 12$ is equivalent to $(2x - 3)(x + 4)$;

♦ x^5x^{-2} is equivalent to x^3 as long as $x \neq 0$;

♦ $3x/2000 + 8x/2000$ is equivalent to $11x/2000$.

Students in such algebra classes also learned rules to enable them to produce equivalent equations, or equations with the same roots. Successively applying these rules allowed them to find an equivalent form of an equation or system of equations for which the roots are apparent. For example, by applying such properties of real numbers as the distributive, associative, and commutative properties and such properties of equality as the addition and multiplication properties, students learned to produce the following set of equivalent equations:

$$3(x + 5) = (x + 3)^2 - 4$$
$$3x + 15 = (x + 3)^2 - 4$$
$$3x + 15 = (x^2 + 6x + 9) - 4$$
$$3x + 15 = (x^2 + 6x) + (9 - 4)$$
$$3x + 15 = x^2 + 6x + 5$$
$$0 = x^2 + 3x - 10$$
$$0 = (x + 5)(x - 2)$$

They learned that since $x = -5$ and $x = 2$ satisfy the last equation and since the rest of the equations are equivalent, $x = -5$ and $x = 2$ must also satisfy each of them. They learned parallel rules and procedures for inequalities.

SYMBOLIC MANIPULATION IN A TECHNOLOGICAL WORLD

Armed with today's technology, rather than rewriting expressions without apparent purpose, students must understand how the information conveyed by the expression is altered in its different forms. For example, it is easy to approximate the value of $n = 10/\sqrt{99}$. The value is a little larger than 1, since $\sqrt{99}$ is a little less than 10. The approximation process is not as easy if n is given in its equivalent form, $10\sqrt{11}/33$. Similarly, although

$$\frac{1}{a} + \frac{2}{b} + \frac{3}{c} = \frac{b \cdot c + 2 \cdot a \cdot c + 3 \cdot a \cdot b}{a \cdot b \cdot c},$$

no student with a calculator would choose to evaluate the second expression rather than the first when given the values of *a, b,* and *c.* Although the second expression is billed as "simplified," it is not at all clear that the ten-operation expression on the right is "simpler" than the five-operation expression on the left.

In a technological world, it is still useful to produce equivalent symbolic forms of expressions and to solve equations. The difference is that these equivalent forms and equation solutions can be produced automatically through computer-algebra programs or graphing and symbolic calculators.

With computers and calculators available to produce equivalent expressions and to solve equations, students can concentrate on the more global aspects of work with symbols. What algebra students of the future will learn about symbols will differ greatly from what their present-day counterparts are learning. Three ways in which symbolic work will differ follow:

♦ There will be a greater concentration on learning the meaning of symbols as representations.

♦ Instead of focusing only on the application of rules for producing equivalent forms, work with equivalent forms will center on understanding the meaning of equivalence.

♦ Work in solving equations will draw on new ways of viewing the properties of equality and will involve greater attention to interpreting the solutions produced by computer-algebra systems.

The Meaning of Symbols as Representations

Currently, in spite of the best efforts of their teachers, many students come away from their algebra experience with a fragile understanding of the concepts and procedures with which they have been working. In particular, they come to view symbolic representations as entities to be manipulated rather than as meaningful representations in themselves.

As a rule, the manipulations being taught proceed from "unsimplified" to "simplified" form. Students seldom get an opportunity to ask about the ultimate purpose of the expression. As just seen, the "unsimplified" form may be more useful for a particular task. Before manipulating an expression, students should question which possible form offers the most useful information. Clearly, a technological algebra in which the present fixed hierarchy of form is questioned will require a more reflective and thoughtful approach by students. Understanding will be far more important than remembering.

When teachers capitalize on technology, they can teach algebra so that students enhance their understanding through symbolic, graphical, and numerical representations as well as through the interrelationships among these representations.

Direct Information from Symbolic Representations

At times, symbolic representations give information directly.

1. The symbolic representation $k(c) = 2c + 3$ suggests a linear function *k* whose graph has slope 2 and intersects the output axis at 3 (fig. 6.1).

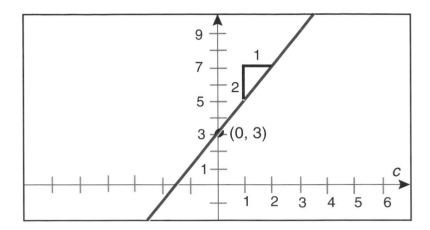

Fig. 6.1. Graph of k(c) = 2c + 3

2. A factored symbolic form can reveal the zeros of a function. The symbolic representation $g(a) = (a - 3)(a + 7)$ suggests that the graph of g intersects the a-axis at 3 and -7 (fig. 6.2).

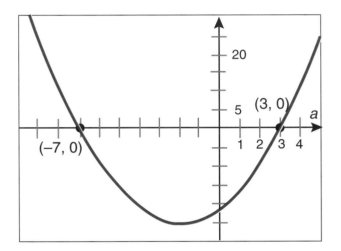

Fig. 6.2. Graph of g(a) = (a − 3)(a + 7)

3. Symbolic representations can reveal the "end behavior" for functions. For example, in the function

$$f(x) = \frac{10}{x} + 3x + 5,$$

the term

$$\frac{10}{x}$$

contributes very little to the value of $f(x)$ when $|x|$ is very large. That is, the symbolic representation

$$f(x) = \frac{10}{x} + 3x + 5$$

suggests a graph that approximates the graph of $d(x) = 3x + 5$ for very large and very small values of x (fig. 6.3).

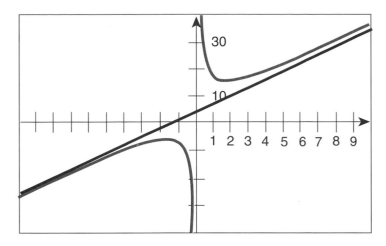

Fig. 6.3. Graphs of f(x) = 10/x + 3x + 5 and d(x) = 3x + 5

A similar argument yields information about *f*(*x*) when |*x*| is sufficiently small. That is, the symbolic representation *f*(*x*) = 10/*x* + 3*x* + 5 suggests a graph that approximates the graph of *h*(*x*)= 10/*x* for values of *x* very close to 0 (fig. 6.4).

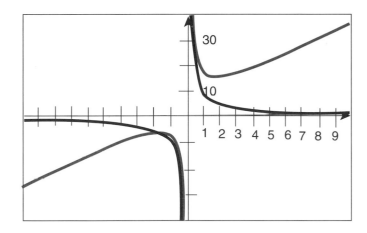

Fig. 6.4. Graphs of f(x) = 10/x + 3x + 5 and h(x) = 10/x

4. Different symbolic forms are often useful in applied settings because they give important information about relationships among quantities. One example arises in the classic problem of minimizing the surface area of a can. For a can with a volume of 355 cc (a soft-drink can) whose shape is modeled by a right circular cylinder, the surface area is given by

$$A(r) = 2\pi r^2 + 710/r$$

where *r* is the radius of the can in centimeters. The function rule *A*(*r*) is the sum of a quadratic function, $A_q(r) = 2\pi r^2$, which gives the contributions of the top and the bottom to the total area, and a rational function, $A_r(r) = 710/r$, which gives the contribution of the side of the can to the total area. Studying the graphs of these two component functions (fig. 6.5) over the restricted domain of reality allows the students to "see" more clearly the presence of a minimum value in their sum.

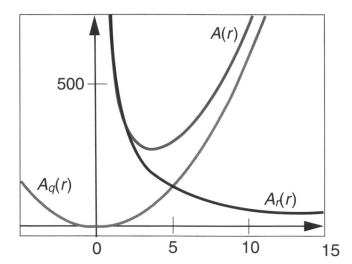

Fig. 6.5. Graph of A(r) and the graphs of its component function rules, $A_q(r)$ and $A_r(r)$

The symbolic and graphical representations for these functions are useful in interpreting the quantitative relationships.

The behavior of the sum function is reflected in the behavior of the two component functions. Since the value of the quadratic is small for r values close to 0, the sum will be only a little larger than the value of the rational function for small values of r. Similarly, since the value of the rational function is small for large values of r, then the sum will be only a little larger than the value of the quadratic for large values of r. So, the sum function will be close to the quadratic curve at certain intervals of the domain and close to the rational curve at other points. In terms of the model, this argument tells the student that the top and bottom of the can add little to the total surface area when r is small but compose most of the area when r is large.

As r increases, the value of A will be decreasing with A_r and increasing with A_q. Because A_r and A_q are monotonic and continuous, the sum function must change from being a decreasing function to being an increasing function somewhere at a minimum value.

Here, the "simplest" form is not the most interpretable. If we "simplify" the function by rewriting it as

$$A(r) = 2\left(\frac{\pi r^3 + 355}{r}\right),$$

it becomes more difficult to picture the function. We also lose the physical interpretation of the components of the total surface area generated by the top and bottom and by the side.

Access to a Greater Variety of Forms

Chapter 3, on families of functions, argues for the importance of students' being able to recognize the relationship between the shape of a curve and its symbolic form. For example, a linear shape implies a function of the form $f(x) = ax + b$; a parabolic shape should be associated with the function $g(x) = ax^2 + bx + c$; and exponential functions of the form $h(x) = a^x$ should be matched with the classic exponential-growth curve. These forms, however, are not the only ones that these functions can take.

Technology offers a new way of thinking about and writing these functions: recursive equations. In the future, when students see a linear shape, they should think about both $f(x) = ax + b$ and $y_0 = b$, $y_n = y_{n-1} + m$. Both forms represent this linear function. Similarly, parabolic shapes should elicit thoughts of $g(x) = ax^2$ and $y_n = y_{n-1} + a(2n - 1)$, and exponential curves should bring both $h(x) = a^x$ and $y_n = ay_{n-1}$ to mind. Technology gives us more ways to think about functions, each with its own representation.

If we think about the classic compound-interest problem, we have $A_0 = 1000$ and either $A_n = A_{n-1} + 0.05A_{n-1}$ or the more compact form, $A_n = 1.05A_{n-1}$. Each form has its own use. The first recursive equation describes the two parts of the total money in the bank—what was there before and left untouched and the new interest acquired. The second, more compact form allows for efficient computations.

Equivalent Expressions

A major part of traditional algebra curricula centers on "simplifying" algebraic expressions. In other words, algebra classes typically focus on producing particular types of equivalent forms for algebraic expressions—equivalent forms that conform to some agreed-on canon. Students too often leave their algebra experience with a modicum of ability to produce equivalent forms but very little understanding of the meaning of that equivalence. In a technological world in which students have access to computer-algebra utilities, including graphers, spreadsheets, and symbolic-manipulation programs, the importance of producing equivalent forms need no longer overshadow the importance of understanding what the equivalent expressions mean.

Consider, for example, four different but equivalent representations:

$$h\left(\frac{a+b}{2}\right)$$

$$\frac{1}{2}ha + \frac{1}{2}hb$$

$$\frac{1}{2}h(b - a - d) + ha + \frac{1}{2}hd$$

$$hb - \frac{1}{2}h(b - a)$$

These expressions are all representations for the area of the trapezoid in figure 6.6, and each is tied to a different way of computing the area. The first expression,

$$h\left(\frac{a+b}{2}\right),$$

suggests that the area of a trapezoid is the product of the height and the average (arithmetic mean) of the two bases. (This conceptualization of the area could also describe the area of a rectangle and the area of a parallelogram.)

Try This: **To introduce the notion of the equivalence of algebraic expressions, ask students to use the pattern suggested in the following sequence of figures to come up with a symbolic rule for the total number of toothpicks as a function of the number of toothpicks in the bottom row. Different counting methods will result in different appropriate symbolic rules, all of which will be equivalent. This activity can lead into a discussion on the equivalence of algebraic expressions.**

The following design is made up of toothpicks of equal length. The first few figures are shown; the last figure has four toothpicks across the bottom, and the pattern can be extended to any number of rows. How many toothpicks are needed if a total of ten toothpicks is across the bottom? A total of n toothpicks?

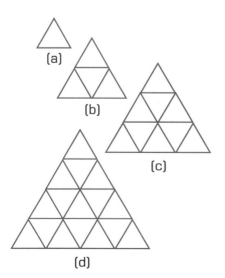

(a)

(b)

(c)

(d)

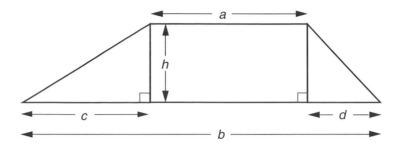

Fig. 6.6. Area of trapezoid = h(a + b)/2.

A different way of computing the area of the trapezoid involves dissecting the trapezoid into two triangles (fig. 6.7) and results in the equivalent expression

$$\frac{1}{2}ha + \frac{1}{2}hb.$$

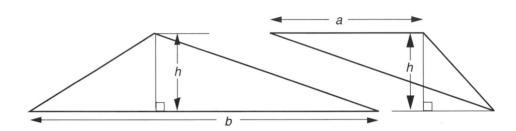

Fig. 6.7. Area of trapezoid = (1/2)ha + (1/2)hb.

A still different dissection yields a different equivalent expression (fig. 6.8),

$$\frac{1}{2}h(b - a - d) + ha + \frac{1}{2}hd.$$

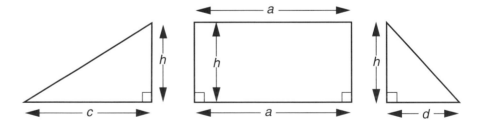

Fig. 6.8. Area of trapezoid = (1/2)hc + ha + (1/2)hd =
(1/2)h(b – a – d) + ha + (1/2)hd.

These last two representations came from dissecting the trapezoid into other figures. Still another equivalent form results from subtracting the areas of familiarly shaped regions. In this example, the area of the trapezoid can be conceived of as the area of a rectangle minus the areas of two triangles, as shown in figure 6.9. Here, the area of the trapezoid is the area, *hb,* of a rectangle minus the areas of triangles 1 and 2.

A rearrangement of these figures yields a slightly less complicated expression. Instead of subtracting the areas of triangle 1 and triangle 2 separately, we can "move" the triangle with area 2 to the left of triangle 1,

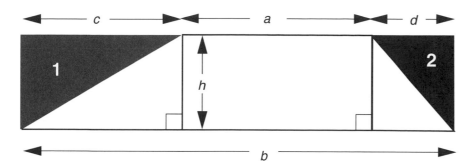

Fig. 6.9. Area of trapezoid = hb – area of triangle 1 – area of triangle 2.

combining the two triangular regions into a single triangular region. The area of the trapezoid is the area of the resulting parallelogram minus the area of a triangle with base $c + d$ (fig. 6.10).

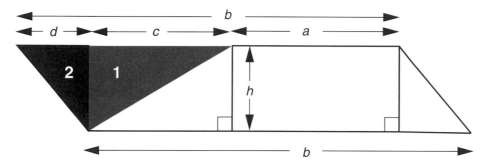

Fig. 6.10. Area of trapezoid = hb – area of new triangle = hb – (1/2)h(d + c). Since d + c = b – a, the area = hb – (1/2)h(b – a).

These four algebraic expressions,

$$h\left(\frac{a+b}{2}\right),$$

$$\frac{1}{2}ha + \frac{1}{2}hb,$$

$$\frac{1}{2}h(b - a - d) + ha + \frac{1}{2}hd,$$

and

$$hb - \frac{1}{2}h(b - a),$$

are equivalent because they all accurately represent the area of the same trapezoid. The equivalence of these expressions can be corroborated numerically and symbolically. Numerically, for the same input values, each expression gives the same output value. Symbolically, all four can be seen to be equivalent because they are all equivalent to the same symbolic expression, $(1/2)ha + (1/2)hb$, or to any of the three other expressions.

*Teaching Matters: Using geometric figures to represent what are today called "algebraic ideas" is the core of the ancient Greeks' geometrical algebra in which they actually solved equations using geometric constructions. Have students explore the history of this topic and write a report of their findings. Most histories of mathematics include sections on Greek geometric algebra. A good source is **Historical Topics for the Mathematics Classroom (NCTM 1989).***

Equivalent Equations

In the copying-machine situation in chapter 1, we saw the graphical effect of subtracting the same constant from both sides of an equation. Now consider solving the equation

$$x^2 + 2x = 5 - 2x.$$

Graphically, its solution is the set of x-values for the intersection of the functions f and g where $f(x) = x^2 + 2x$ and $g(x) = 5 - 2x$ (see fig. 6.11).

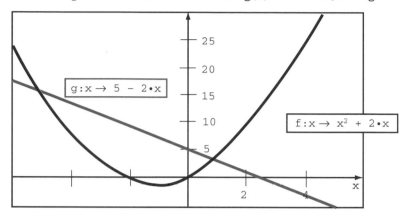

Fig. 6.11. The simultaneous graphs of $f(x) = x^2 + 2x$ and $g(x) = 5 - 2x$ gives a graphical representation of the solution of $x^2 + 2x = 5 - 2x$.

Figure 6.12 suggests that the x-values of the intersection of $f(x) = x^2 + 2x$ and $g(x) = 5 - 2x$ are the same as the x-values of the intersection of $h(x) = x^2 + 2x - 5$ and $k(x) = -2x$. This is not surprising in light of the fact that subtracting 5 from the output values of a function will shift the graphs of each original function down five units. Thus, adding a constant to both sides of an equation produces an equivalent equation.

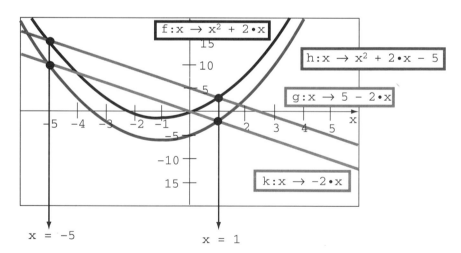

Fig. 6.12. The simultaneous graphs of $f(x) = x^2 + 2x$ and $g(x) = 5 - 2x$ give a graphical representation of the solution of $x^2 + 2x = 5 - 2x$. The simultaneous graphs of $h(x) = x^2 + 2x - 5$ and $k(x) = -2x$ give a graphical representation of the solution of $x^2 + 2x - 5 = -2x$. The diagram shows that the solutions of these two equations are the same.

It is slightly more complicated to think about the likely effect of subtracting a variable expression from both sides of an equation. Figure 6.13 gives a graphical illustration of the fact that the solutions for $2x + 13 = 5x + 4$ are the same as the solutions for $13 = 3x + 4$. Here, the slopes of the two linear functions decreased by the same number, so the slope of 2 in $f(x) = 2x + 13$ decreased to a slope of 0 in $h(x) = 13$, and the slope of 5 in $g(x) = 5x + 4$ decreased to a slope of 3 in $k(x) = 3x + 4$. Even with these shifts in slope, the x-value, 3, of the point of intersection of f and g is the same as the x-value of the point of intersection of h and k.

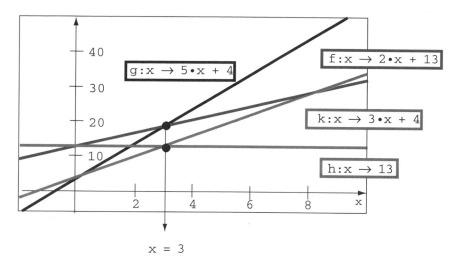

x = 3

Fig. 6.13. The simultaneous graphs of f(x) = 2x + 13 and g(x) = 5x + 4 give a graphical representation of the solution of 2x + 13 = 5x + 4. The simultaneous graphs of h(x) = 13 and k(x) = 3x + 4 give a graphical representation of the solution of 13 = 3x + 4. The diagram shows that the solutions of these two equations are the same.

For each property of addition and the multiplication properties of equality and inequality, similar graphical interpretations can be constructed. Although it is informative to explore the graphical meaning of such transformations, it is clearly not expedient to construct such graphical representations for each transformation in the solution procedure. A more complete development of this graphical interpretation of properties of equality is contained in chapter 9 of *Concepts in Algebra: A Technological Approach* (Fey and Heid 1995).

SYMBOLIC REASONING IN ANALYZING REAL-WORLD SITUATIONS

The technology that has already had the greatest impact on the algebra taught in schools is the technology of visualization and, in particular, easily accessible graphing technology. Spreadsheets and other data-analysis tools also have influenced current thinking about school algebra. The technology most likely to have additional influence in the future is convenient and user-friendly computerized symbolic manipulation. With graphing utilities, students have enhanced their mathematical understanding by representing mathematical functions and relations in new visual ways. With symbolic-manipulation tools, students will be able to deepen their understanding of the familiar symbolic representation while making new connections among symbolic, numerical, and graphical representations.

The role of symbolic reasoning becomes more evident in the context of rich, applied settings. The following section, which addresses the role of symbolic reasoning in a problem situation and related items, is adapted from *Concepts in Algebra: A Technological Approach* (Fey and Heid 1995). The Teaching Matters notes comment on the role of symbol sense or symbolic reasoning in the solution process.

T-Shirt Situation

Each year your school has a campaign to sell school T-shirts. A tradition at the school is that T-shirts will be sold only to students and that each student is entitled to buy only one T-shirt. Suppose that your task is to order a supply of shirts and set the price. Your goal is to maximize profit for the project, since the profit money will go toward the cost of a class trip.

Estimate the demand function for the T-shirts. Survey your class to find out how many shirts would be purchased at each of the following prices. Then use the actual number of students in your school to estimate the number of shirts that would be sold at each price.

Price in dollars	Number of students in your class who would buy at this price	Fraction of your class who would buy at this price	Number of students in the whole school who would buy at this price
4	_____	_____	_____
8	_____	_____	_____
12	_____	_____	_____
16	_____	_____	_____
20	_____	_____	_____

Find a demand function s that predicts sales of T-shirts as a function of price p.

Here, students in a school with a population of 1590 students collected the following data from their classmates, asking them if they would buy a T-shirt at each of the following prices:

Name	$4	$8	$12	$16	$20
Joshua	Yes	Yes	Yes	No	No
Michael	Yes	Yes	Yes	Yes	No
Sally	No	No	No	No	No
Bruce	Yes	No	No	No	No
Tina	Yes	No	No	No	No
Istvan	Yes	Yes	Yes	Yes	No
Rod	Yes	Yes	No	No	No
Katrina	Yes	No	No	No	No
Shelly	Yes	Yes	No	No	No
Pedro	Yes	No	No	No	No
Tawanna	Yes	Yes	Yes	Yes	No
Hashem	Yes	Yes	Yes	Yes	No
Juan	Yes	Yes	No	No	No
Jeff	No	No	No	No	No
Maria	Yes	Yes	Yes	No	No
Billie	Yes	Yes	Yes	Yes	Yes
Nathan	Yes	Yes	No	No	No
Barry	Yes	Yes	No	No	No
Gabor	Yes	Yes	Yes	Yes	No
George	Yes	Yes	Yes	No	No

Jose	Yes	Yes	No	No	No
Svetlana	Yes	Yes	Yes	Yes	No
Juanita	Yes	Yes	Yes	No	No
Sylvia	Yes	No	No	No	No
Sarah	Yes	Yes	No	No	No
Penelope	Yes	No	No	No	No
Jennifer	Yes	Yes	No	No	No
Raoul	Yes	No	No	No	No
Jean	Yes	Yes	Yes	Yes	Yes
Tony	Yes	No	No	No	No

As could have been predicted, once a classmate had said no to a price, he or she was never willing to buy the T-shirt for a higher price. The data from this survey are summarized in the following table.

Price of a T-shirt	Number of students in your class who would buy at this price	Fraction of your class who would buy at this price	Number of students in the whole school who would buy at this price
4	28	$\frac{28}{30}$	1484
8	20	$\frac{20}{30}$	1060
12	12	$\frac{12}{30}$	636
16	8	$\frac{8}{30}$	424
20	2	$\frac{2}{30}$	106

At this point, students may graph the number of T-shirts that will be sold as a function of the price at which they are sold (see fig. 6.14).

InData	OutData
4	1484
8	1060
12	636
16	424
20	106

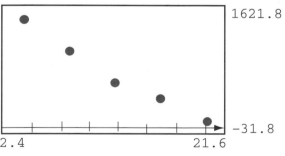

Fig. 6.14. Graph and table depicting the number of T-shirts expected to be sold as a function of the price

Students might construct a table of function values for the demand function, graph the corresponding ordered pairs, and fit a curve to those data. Accomplishing this task usually requires students to specify the family of functions to which the data are being fit and so calls on the students' knowledge of the relationship between a family of functions and its shape. Students who connect the shape of a line with its symbolic representation may recognize a general downward linear trend and choose a linear fit. The computer would then generate a linear rule like the one in figure 6.15.

InData	OutData
4	1484
8	1060
12	636
16	424
20	106

$$F1{:}x \rightarrow -\frac{424}{5} \cdot x + \frac{8798}{5}$$

Fig. 6.15. Graph, table, and linear rule depicting the number of T-shirts expected to be sold as a function of the price. The rule was generated by a technological curve fitter.

After generating the symbolic rule, students might examine the function rule or its equivalent, $F1(x) = -84.8x + 1759.6$, to decide if the rule seems reasonable. Students might notice that for every dollar increase in the price of a T-shirt, about eighty-four fewer shirts will be sold. This might seem to be a plausible condition. The function rule also suggests that about 1760 could be given away, or "sold" at a price of $0. This fact would seem unreasonable to students who recall the "only one T-shirt for each student" rule.

Find a rule that predicts revenue R as a function of price.

Students might first construct a table of function values for the revenue function, graph the corresponding ordered pairs, and fit a curve to those data. In this example, students might notice a general parabolic pattern, make the graphical-symbolic connection between a parabolic shape and a quadratic, and fit a quadratic as shown in figure 6.16.

InData	OutData
4	5936
8	8480
12	7632
16	6784
20	2120

$$F2{:}x \rightarrow -\frac{901}{14} \cdot x^2 + \frac{45898}{35} \cdot x + \frac{8904}{5}$$

Fig. 6.16. Graph, table, and quadratic rule depicting the revenue expected from T-shirt sales as a function of the price. The rule was generated by a technological curve fitter.

◆　　◆　　◆　　◆　　◆　　◆　　◆　　◆

Alternatively, students might take a somewhat more symbolic approach and formulate a function rule for revenue on the basis of the linear rule that fits the demand data. Here, the rule would be

$$R_1(x) = x\left(-\frac{424}{5}x + \frac{8\,798}{5}\right).$$

Formulating a rule in this way seems reasonable, since the rule is the product of two linear functions, the product of a linear and a linear is quadratic, and the graph of the relationship looks quadratic. With slightly different symbolic reasoning, however, students may conclude that the two rules for revenue,

$$R_1(x) = x\left(-\frac{424}{5}x + \frac{8\,798}{5}\right)$$

and

$$R_2(x) = -\frac{901}{14}x^2 + \frac{45\,898}{35}x + \frac{8\,904}{5},$$

are not equivalent. Students may notice that although the two rules are both quadratic, the zeros of the function rules are different. They may create and compare graphs and tables as shown in figure 6.17.

$$R1: x \rightarrow -\frac{424}{5} \cdot x^2 + \frac{8798}{5} \cdot x$$

$$R2: x \rightarrow -\frac{901}{14} \cdot x^2 + \frac{45898}{35} \cdot x + \frac{8904}{5}$$

x	R1 (x)	R2 (x)
4	5681.60	5996.57
8	8649.60	8152.91
12	8904	8249.83
16	6444.80	6287.31
20	1272	2265.37

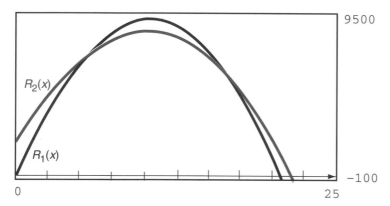

Fig. 6.17. Graph, table, and two quadratic rules depicting the revenue expected from T-shirt sales as a function of the price. The rule $R_1(x)$ was derived from the curve-fitted demand function previously derived; the rule $R_2(x)$ was generated by a technological curve fitter from data generated using the curve-fitted demand function.

Teaching Matters: Symbolic reasoning suggests that the two function rules are not equivalent. Although they are both quadratic functions, they have at least one different parameter.

An important category of symbol sense is the ability to reason about (though not necessarily execute by hand) symbolic procedures or algorithms. In the T-shirt example, two different rules were derived by applying the same algorithm (least-squares curve fitting) at different junctures.

Teaching Matters: A reasoned analysis of real-world situations requires moving between and among graphical, numerical, and symbolic representations of the quantitative relations in the situation. Symbolic reasoning links symbolic representations with graphical, contextual, or numerical representations.

When computing tools become a regular part of mathematical explorations, it is necessary to continue to monitor the reasonableness of results. A consciousness of the meaning of the algorithms being used is helpful.

A glance at the tables and graphs reveals that the two function rules give revenue values that are close but not equal. Although both rules legitimately model the same quantitative relation, they are not equivalent. What is the significance of the y-intercepts of the two models? Which seems more realistic?

What facts about the procedures that generated the rules made them different? Both involved curve fitting—one a linear fit of the demand data and the other an additional quadratic fit of the related revenue data. Depending on the readiness of the students, these results might generate a discussion about the numerical consequences of using curve fitting and of using computing tools to generate curves of best fit.

> Find the price that gives maximum revenue. Then calculate the number of T-shirts you would expect to sell at this price.

Since both symbolic rules are quadratic and the coefficient of the quadratic term in each is negative, each of these functions must have a maximum. Thus, symbolic evidence supports the graphical evidence suggesting maxima. The preceding table indicates that the maximum for each function has input value that can be found in the interval (8, 16). From the graph it is clear that the maximum for each function occurs between 7.5 and 12.5, that is, between the fourth and sixth tick marks on the horizontal axis. Combining these two facts makes it seem reasonable to begin the search for maxima with x-values ranging between 8 and 12.5.

Before launching into a scan-and-zoom table routine, students might decide that the nearest quarter is "close enough." Successive tables "zooming in" on the maximum value suggest that the maxima are actually distinct in spite of their appearing to coincide on the initial graphs.

x	$R_1(x)$	$R_2(x)$
8.0	8649.6	8152.9
8.5	8829.8	8277.7
9.0	8967.6	8370.2
9.5	9063.0	8430.6
10.0	9116.0	8458.8
10.5	9126.6	8454.8
11.0	9094.8	8418.7
11.5	9020.6	8350.3
12.0	8904.0	8249.8
12.5	8745.0	8117.1

This first table suggests that the maximum for R_1 occurs between 10 and 11 and the maximum for R_2 occurs between 9.5 and 10.5. A reasonable next table might examine input-output pairs between 9.5 and 11.

x	$R_1(x)$	$R_2(x)$
9.50	9063.00	8430.60
9.75	9094.80	8448.72
10.00	9116.00	8458.80
10.25	9126.60	8460.83
10.50	9126.60	8454.83
10.75	9116.00	8440.77
11.00	9094.80	8418.67

This table suggests that when T-shirt prices are confined to multiples of a quarter, the maximum for R_1 occurs at 10.25 and 10.50 and the maximum for R_2 occurs at 10.25. If we have no particular reason to prefer one model to the other, we might choose the lower price of $10.25 a shirt.

The company that will make the T-shirts quotes the following cost data:

Number of Shirts	Total Cost in Dollars
100	500
250	1100
500	2000
750	2750
1000	3450
1500	4950
2000	6000

Find a rule that gives total cost as a function of the number of shirts ordered.

Plotting the ordered pairs of input-output data, noting a linear trend, and fitting a linear function, students direct the computing tool to generate a linear-function fit, as shown in figure 6.18.

Teaching Matters: In fitting this curve, students display symbol sense in knowing how symbolic and graphical representations are connected __and__ in making the symbol-graph connection at appropriate times.

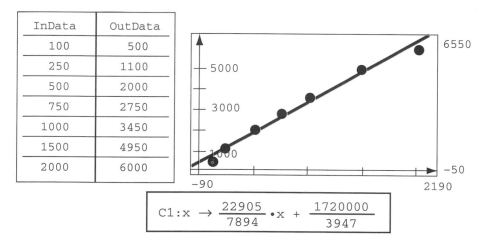

InData	OutData
100	500
250	1100
500	2000
750	2750
1000	3450
1500	4950
2000	6000

$$C1:x \rightarrow \frac{22905}{7894} \cdot x + \frac{1720000}{3947}$$

Figure 6.18. Graph, table, and linear rule depicting the total cost expected to accrue from T-shirt sales as a function of the number of T-shirts sold. The linear rule was generated by a technological curve fitter.

Checking again to determine if the rule makes sense, students might first round coefficients and constants to two decimal places and get $C_1(x) = 2.90x + 435.77$. This suggests a set-up cost of about \$435.77 and an increase in total cost of about \$2.90 for each additional shirt bought or produced. For some printing processes, the set-up cost might seem reasonable, especially if it includes the cost of creating the design for the T-shirt. The \$2.90 variable cost seems very inexpensive and might be a reflection of poor T-shirt quality.

Teaching Matters: Symbolic reasoning here involves interpreting symbolic expressions in applied situations and determining whether these expressions make sense.

Checking again to determine if the rule makes sense, students might first round coefficients and constants to two decimal places and get $C_1(x) = 2.90x + 435.77$. This suggests a set-up cost of about \$435.77 and an increase in total cost of about \$2.90 for each additional shirt bought or produced. For some printing processes, the set-up cost might seem reasonable, especially if it includes the cost of creating the design for the T-shirt. The \$2.90 variable cost seems very inexpensive and might be a reflection of poor T-shirt quality.

> Find a rule for the cost as a function of the selling price of each shirt.

Teaching Matters: Students with symbol sense understand the relevance of the composition of functions. Here, because cost is a function of the number of shirts sold and the number of shirts sold is a function of the price of the shirt, then through the composition of functions, cost can be expressed as a function of the price of a shirt. Symbol sense requires the continual assessment of both mathematical and contextual reasons for a symbolic expression to make sense.

To find the cost as a function of the price of each shirt, students can combine the function rules for cost as a function of the number of shirts sold and the number of shirts sold as a function of the price of each shirt. That is, they can combine $C_1(x) = 2.90x + 435.77$ with $n(p) = -84.8p + 1759.6$. So $C_1(n(p)) = -245.92p + 5538.61$. See figure 6.19 for an example of using a symbolic-manipulation program to compose these functions. We can say that cost is a function C_2 of price p with rule $C_2(p) = -245.92p + 5538.61$. In this example, the cost of producing the T-shirts is \$5538.61 if the T-shirts are given away, and for every dollar increase in the price charged for a T-shirt, the total cost goes down \$245.92.

Command given to Calc T/L II	Response from Calc T/L II
• Define C1 by C1(x) = 2.90*x + 435.77.	C1:x->2.90*x + 435.77
• Define n by n(p) = -84.8*p + 1759.6.	n:p->-84.8*p + 1759.6
• Write C1(n(p)).	C1(n(p))
• Compute (symbolically).	C1(n(p)) = -245.920*p + 5538.61

Fig. 6.19. Commands given to a symbolic-manipulation program (Calc T/L II) to create the composition of two functions and the computer-generated responses

This function rule for total cost as a function of price charged for each shirt makes symbolic sense for several reasons. First, an abstract mathematical reason gives the rule symbolic sense. The composition of a linear function, C_1, with a linear function, n, is, again, a linear function. Second, the symbolic rule makes sense within the context of the situation. The \$5538.61 cost of producing the T-shirts if the T-shirts are given away makes sense in light of the prediction that about 1760 shirts could be distributed if the price were \$0 and the set-up cost were \$435.77. If 1760 shirts are given away, it would cost \$5102.84 to supply them; adding this supply cost to the set-up costs accounts for the \$5538.61 total cost of producing the shirts to give away.

> Find a rule giving the profit *PR* as a function of the selling price of each shirt.

One way to think about the profit is total revenue minus total cost. Since at least two function rules for total revenue are possible, at least two function rules for total profit are possible. We can label the function rules P_1 and P_2 to correspond to the revenue function rules R_1 and R_2. Figure 6.20 shows one use of a symbolic-manipulation program to generate these rules. The coefficients in the function rules in figure 6.20 are decimal approximations; rational coefficients would yield similar, but not necessarily equivalent, results.

Teaching Matters: Symbolic reasoning requires knowing when it makes sense to combine known symbolic expressions to produce a related one. Here, revenue and cost function rules are subtracted to obtain a prof· it function rule.

List of computer commands

- Define n by n(p) = -84.8*p+1759.6.

- Define C1 by C1(x) = 2.90*x+435.77.

- Define C2 by C2(p) = C1(n(p)).

- Define R1 by R1(p) = p*n(p).

- Define R2 by R2(p) = -64.36*p^2+1311.37*p+1780.80.

- Define P1 by P1(p) = R1(p)-C2(p).

- Simplify image element.

- Define P2 by P2(p) = R2(p)-C2(p).

List of functions generated by this set of commands

| n:p → -84.8•p + 1759.6 |
| C1:x → 2.90•x + 435.77 |
| C2:p → -245.920•p + 5538.61 |
| R1:p → p•(-84.8•p + 1759.6) |
| R2:p → -64.36•p² + 1311.37•p + 1780.80 |
| P1:p → p•(-84.8•p + 1759.6) + 245.920•p - 5538.61 |
| SP1:p → -84.8•p² + 2005.52•p - 5538.61 |
| P2:p → -64.36•p² + 1557.29•p - 3757.81 |

Fig. 6.20. The list of function rules was generated using a symbolic-manipulation program (Calc T/L II) and the indicated list of computer commands.

Find the price that will guarantee a break-even result (profit is 0), and find the predicted sales at that price (fig. 6.21).

The profit function P_1 (used with revenue function R_1) suggests that the profit would be 0 if the price of each shirt were about \$3.19 or \$20.46. The "nearest quarter" condition suggests a price of \$3.25 or \$20.50 to break even. Since we have the function rules available, we will analyze the situation for prices surrounding \$3.19 and \$20.46.

```
┌──────────────────────────────────────────────────┐
│  ┌─────────────────────────────────────────┐     │
│  │ P1:p  →  -84.8•p² + 2005.52•p - 5538.61  │     │
│  └─────────────────────────────────────────┘     │
│                                                    │
│  P1(p) = 0   if p is in  {p1 , p2} where   ⎛ p1 = 3.19268 ⎞ │
│                                            ⎝              ⎠ │
│  ⎛ p2 = 20.4573 ⎞                                 │
│  ⎝              ⎠                                 │
│  ┌─────────────────────────────────────────┐     │
│  │ P2:p  →  -64.36•p² + 1557.29•p - 3757.81 │     │
│  └─────────────────────────────────────────┘     │
│  P2(p) = 0     if p is in  {p3 , p4} where  ⎛ p3 = 21.4780 ⎞ │
│                                             ⎝              ⎠ │
│  ⎛ p4 = 2.71846 ⎞                                 │
│  ⎝              ⎠                                 │
└──────────────────────────────────────────────────┘
```

Fig. 6.21

We can use a symbolic manipulator to produce the following results:

$$n(3.25) = 1484$$

$$P_1(3.25) = 83.63$$

$$n(3.00) = 1505$$

$$P_1(3.00) = -285.25$$

$$n(20.25) = 42$$

$$P_1(20.25) = 299.87$$

$$n(20.50) = 21$$

$$P_1(20.50) = -63.00$$

If we use P_1 as the profit function and insist on charging an amount rounded to the nearest quarter, the closest we can get to "breaking even" is to set the price at \$3.25 and expect a profit of about \$84.00 or set a price of \$20.50 and expect a loss of about \$63.00.

The profit function P_2 (used with revenue function R_2) suggests that the profit would be \$0 if the price of each shirt were about \$2.72 or \$21.48. Since we have the function rules available, we will analyze the situation for prices surrounding \$2.72 and \$21.48:

$$n(2.75) = 1484$$

$$P_2(2.75) = 38.02$$

$$n(2.50) = 1505$$

$$P_2(2.50) = -266.84$$

$$n(21.25) = 42$$

$$P_2(21.25) = 272.04$$

$$n(21.50) = 21$$

$$P_2(21.50) = -26.49$$

Teaching Matters: Different symbolic rules produce different numerical results. One task involved in symbolic reasoning is keeping track of the origin of each different rule. Keeping the situation conditions in mind helps students interpret the symbolic solutions sensibly.

If we use P_2 as the profit function and insist on charging an amount rounded to the nearest quarter, the closest we can get to "breaking even" is to set the price at \$2.75 and expect a profit of about \$38.00 or set a price of \$21.50 and expect a loss of about \$27.00.

> Find the price to charge to make the maximum profit.

Both profit functions P_1 and P_2 are quadratic. Because of the symmetry of quadratic functions, the price that generates a maximum profit is half-way between the break-even points. Thus, for P_1 the price (to the nearest quarter) that generates a maximum profit is $11.75, whereas for P_2 the price (to the nearest quarter) that generates a maximum profit is $12.00.

Teaching Matters: Symbolic reasoning often involves reasoning that crosses several representations. In this example, the reasoning starts with the symbolic, draws on the graphical, and returns to the numerical.

> Compare the price that maximizes profit with the price that maximizes revenue.

As shown in figure 6.22, the revenue is predicted to be $0 when either $0 or $20.75 is charged. Using a symmetry argument suggests a maximum revenue (to the nearest quarter) when either $10.25 or $10.50 is charged. By similar reasoning using R_2, we might predict a maximum revenue (to the nearest quarter) when the price is $10.25. Although a price of $10.25 or $10.50 would maximize revenue, it takes a slightly higher price to maximize profit.

Teaching Matters: An interesting follow-up project that draws significantly on students' symbolic reasoning centers on whether the price generating maximum profit in a business situation is always higher than the price that would generate maximum revenue. Students could explore the effects of different types of demand functions.

R1:p $\rightarrow$ p•(-84.8•p + 1759.6)

R2:p $\rightarrow$ -64.36•p² + 1311.37•p + 1780.80

R1(p) = 0 if p is in {p1 , p2} where   (p1 = 0)

(p2 = 20.7500)

R2(p) = 0 if p is in {p3 , p4} where

(p3 = -1.27783) (p4 = 21.6533)

Fig. 6.22. Solution to the T-shirt zero-revenue problem using symbolic-manipulation technology

A reasoned analysis of real-world situations requires moving between and among graphical, numerical, and symbolic representations of the quantitative relations in the situations. When we study real-world situations algebraically, each connection is important: symbol-symbol, symbol-number, symbol-graph, and symbol-situation. In the example just explored, the following features of symbol sense and symbolic reasoning arose:

Symbol-symbol connection

◆ The recognition that a new rule can be created by finding the product or difference of two known symbolic rules and an awareness of when it is appropriate to do so (profit = revenue – cost; revenue = number sold * price per item, and so on)

◆ The awareness of what symbolic forms are likely to arise as particular families of symbolic forms are added, subtracted, multiplied, and divided (the product of two linear-function rules is likely to be a quadratic-function rule)

Assessment Matters: This extended real-world problem situation involves many rich mathematical ideas applied in connected ways to model the situation. How can we assess the learning that is intended in such complex situations? The traditional testing perspective decomposes the situation into much smaller skills and concepts and then writes short "test items" to assess these isolated ideas. This approach loses entirely the connectedness of the mathematical ideas and the rich, real-world context. Resist the temptation to decompose. Rather, give students similar situations so they can demonstrate their understanding and ability to apply mathematical ideas in such situations. Clearly, this means giving fewer items on a test—maybe just one in a class period—or giving take-home assessments or extended projects.

◆ The ability to reason about a symbolic process or algorithm and recognize its influence on results obtained

◆ A monitoring of the reasonableness of results, including a consciousness of the symbolic process or algorithm used to produce the results

Symbol-number connection

◆ A recognition at relevant times that different function rules produce different results

Graph-symbol connection

◆ The ability to connect a graph of a particular shape with its symbolic form and the ability to recognize when it would be useful to do so

Symbol-situation connection

◆ The ability to use data arising in a situation to determine if a symbolic rule makes sense

◆ The ability to interpret symbolic expressions in applied settings and to determine whether these expressions make sense in the situation

◆ The tendency to keep the situation conditions in mind while interpreting the symbolic solutions

◆ The recognition of when it is appropriate to find and use the composition of two functions or to find and use the product (or sum or difference or quotient or ...) of two known symbolic expressions (if demand is a function of price and revenue is a function of demand, then revenue can be expressed as a function of price through composition of functions)

◆ The ability to make sense of computer-algebra results related to a given situation

CONCLUSION

In the course of this chapter, we have used a variety of related terms: *symbol sense, symbolic manipulation,* and *symbolic reasoning. Symbol sense* suggests an immediate and active awareness of feasible meanings and interpretations of symbolic forms. Symbol sense includes a sensitivity to the likelihood that the results of particular operations or algorithms could take a particular symbolic form. *Symbolic manipulation* refers to the execution of algorithms that take symbols as input and produce other symbols as output. The execution may or may not be technology assisted and may or may not have rich meaning for the executors. *Symbolic reasoning* refers to making inferences about the relationships among symbols and between symbolic representations and graphical, numerical, and contextual representations.

In technologically rich algebra classrooms, by-hand symbolic manipulation will play a minor role. It will be much more important for students to make sense of the symbolic representations and results that they encounter than to master by-hand procedures for producing those representations and results. In this chapter, we have pointed out how thinking about symbolic representations and symbolic manipulations changes in light of the availability of computer-algebra programs, which include symbolic manipulators, function graphers, curve fitters, and table generators. In addition, we have presented an extended example of the role of symbolic reasoning and symbol sense in exploring real-world situations.

Symbolic reasoning takes place in the context of a mathematical world that also includes graphical, numerical, and applied representations. Some features of symbolic reasoning speak to making connections only among the graphical, numerical, and symbolic worlds, and some speak to making connections between the world of symbols and the real world.

In the first part of this chapter, we discussed how a symbolic representation of a function could give information about a graphical representation. Depending on the form of the representation, different features of a function are more salient. For example, in factored form, a polynomial function immediately reveals its zeros; the form

$$f\left(x\right) = \frac{a}{x} + bx + c$$

of the function f reveals its end behavior, whereas the form

$$f\left(x\right) = \frac{a + bx^2 + cx}{x}$$

is better for revealing zeros. If students can use technology to produce different forms of function rules, they will still need to know which form will be useful to them at which time.

The middle portion of the chapter discussed ways to focus algebra on meaning in dealing with symbolic forms. The historical approach to algebra as a sequence of procedures that enable the student to produce equivalent expressions to "simplify" them can profitably be replaced by an approach that concentrates a student's attention on what is meant by the equivalence of two different forms and on when different forms would be useful. Attention to the by-hand solution of equations can be supplanted by helping students understand the meaning of the solutions and the procedures that produce them.

The final portion of the chapter concentrated on developing the role of symbolic reasoning and symbol sense in exploring real-world situations mathematically. In addition to the symbol-graph, symbol-number, symbol-symbol, and symbol-situation features of symbolic reasoning illustrated in this section, symbol sense requires the continual assessment of both mathematical and contextual reasons for a symbolic expression to make sense. Symbol sense suggests an active awareness of the deep structure of an expression or function rule, and symbolic reasoning suggests the ability to make supportable conclusions about the implications of that structure.

FINAL COMMENTS

The technological world in which students and teachers now operate demands a radical transfiguration of algebra in schools. As we enter the twenty-first century, the study of algebra in schools must focus on helping students describe and explain the world around them rather than on developing and refining their execution of by-hand symbolic-manipulation procedures—procedures that are better accomplished through the informed use of computing tools. Using these graphical, numerical, symbolic, and exploratory multirepresentational computer-algebra tools, students can learn to use algebraic ideas to explore the relationships they learn to observe in their world.

The content of algebra in a technological world naturally organizes itself around the concepts of function and families of functions and the process of mathematical modeling. Students who develop an in-depth understanding of families of functions learn to recognize essential features of a function family in real-world situations. They work between and among graphical, numerical, symbolic, and contextual representations to question, verify, disprove, or improve features of their models. In the process, they have multiple opportunities to communicate about such important mathematical ideas as function and mathematical modeling.

Approaches to algebra that capitalize on technological tools in these ways change what teachers and students do in the classroom. Students ask their own questions, pose their own problems, and use computing tools to solve those problems; teachers focus on understanding each student's constructions of mathematical ideas, on facilitating their students' explorations of these ideas, and on opening and encouraging new avenues of oral and written communication about algebraic ideas,

We have suggested ideas about how algebra should change in light of computing tools. We recognize that the study of algebra will undergo a continual transformation guided by substantially new tools and by new insights into how students learn.

Study the following sequence of figures to determine the number of matchsticks required to build similar figures of different sizes.

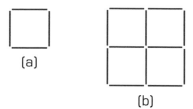

(a)

(b)

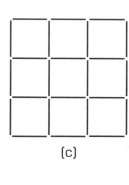

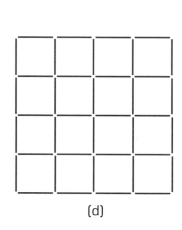

(c)

(d)

1. Use at least two different counting procedures to generate two seemingly different, but equivalent, function rules to describe the number of matchsticks needed as a function *T* of the number of matchsticks, *n*, on one side of the figure. Describe the counting procedure that generated each rule.

2. Offer graphical, numerical, or symbolic evidence to support, but not necessarily prove, your conclusion that the function rules you have constructed are equivalent.

The area of the shaded region in each of the following diagrams can be calculated in several different ways. This activity will engage you in exploring different representations for the indicated areas.

For the first two diagrams, four different procedures for calculating the sum of the areas of the shaded regions are proposed. Write an algebraic expression in terms of the known lengths that matches each procedure. Then determine which procedures are equivalent. Which procedures correctly calculate the total area of the shaded regions?

1. *CB* is the diameter of the circle containing *A*, as shown in the diagram. *ABDE* and *ACGF* are squares. Lengths *AB* and *AC* are assumed to be known.

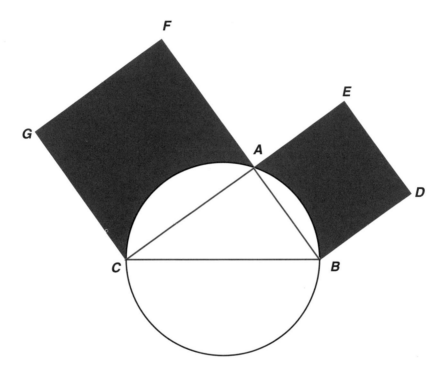

The proposed procedures for determining the sum of the areas of the shaded regions follow.

(*a*) Add the area of trapezoid *BFGC* to the area of square *ABDE*. Then subtract half the area of the circle with diameter $\overline{CB}$. The result is the required area.

(*b*) Add the area of square *ACGF*, the area of square *ABDE*, and the area of triangle *ABC*. Then subtract half the area of the circle with diameter $\overline{CB}$. The result is the required area.

(*c*) Find the area of square *ACGF* and subtract from it the area of the semicircle with diameter $\overline{AC}$. Then find the area of square *AEDB* and subtract from it the area of the semicircle with diameter $\overline{AB}$. The required area is the sum of these two results.

(*d*) Find the area of the two segments of the circle by subtracting the area of triangle *ABC* from the area of the semicircle. Then subtract this result from the sum of the areas of squares *ACGF* and *ABDE*. The result is the required area.

2. *ACMK* is a rectangle. *B, G, L,* and *F* are the mid-points of the sides of *ACMK*. *D, E, H,* and *J* are the midpoints of the sides of quadrilateral *BGLF*. Lengths *AC* and *AK* are assumed to be known.

The proposed procedures for determining the areas of the shaded regions follow.

(*a*) Find the sum of the areas of triangles *BDE, EGJ, JLH,* and *HFD.* The result is the required area.

(*b*) Find the area of rectangle *ACMK* and subtract from it the area of quadrilateral *DEJH.* Subtract this result from the area of quadrilateral *BGLF.* The result is the required area.

(*c*) Find twice the area of triangle *BFG* and subtract from your result the area of quadrilateral *DEJH.* The result is the required area.

(*d*) Find the area of quadrilateral *BGLF* and subtract it from the area of rectangle *ACMK.* Add the area of quadrilateral *BGLF* and the area of quadrilateral *DEJH.* The result is the required area.

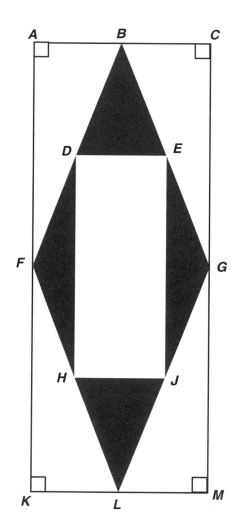

3. For the figure below, find two equivalent expressions for the sum of the areas of the shaded regions.

ABCD is a parallelogram. *J* is the intersection of diagonals $\overline{AC}$ and $\overline{BD}$. *E* is the midpoint of $\overline{JD}$, *F* is the midpoint of $\overline{JA}$, *G* is the midpoint of $\overline{JB}$, and *H* is the midpoint of $\overline{JC}$. $\overline{AD}$ is perpendicular to $\overline{AC}$.

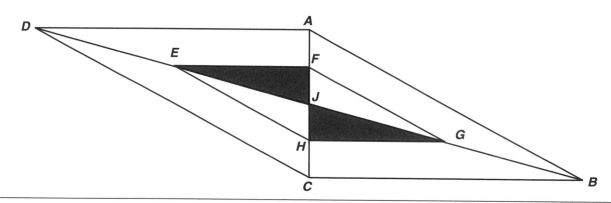

REFERENCES

Blume, Glendon W. "Using the Calculator as an Aid to Concept Development: A Background Unit for Logarithmic and Exponential Functions." *Calculators/Computers Magazine* 3 (January-February 1979): 82–92.

Burrill, Gail, John C. Burrill, Pamela Coffield, Gretchen Davis, Diann Resnick, and Murray Siegel. *Data Analysis and Statistics across the Curriculum. Curriculum and Evaluation Standards for School Mathematics* Addenda Series, Grades 9–12. Reston, Va.: National Council of Teachers of Mathematics, 1992.

Calculus T/L II. Pacific Grove, Calif.: Brooks/Cole Publishing Co., 1990. Software.

Confrey, Jere, et al. Function Probe 2.3.5. Cornell, N.Y.: Cornell Research Foundation, 1992. Software.

Core-Plus Mathematics Project (CPMP). *Linear Models.* Preliminary version. Kalamazoo, Mich.: Western Michigan University, 1993.

Coxford, Arthur F., Linda Burks, Claudia Giamati, and Joyce Jonik. *Geometry from Multiple Perspectives. Curriculum and Evaluation Standards for School Mathematics* Addenda Series, Grades 9–12. Reston, Va.: National Council of Teachers of Mathematics, 1991.

Dugdale, Sharon, Linda J. Wagner, and David Kibbey. "Visualizing Polynomial Functions: New Insights from an Old Method in a New Medium." *Journal of Computers in Mathematics and Science Teaching* 11 (1992):123–41.

Fey, James T., and M. Kathleen Heid, with Richard A. Good, Charlene Sheets, Glendon W. Blume, and Rose Mary Zbiek. *Concepts in Algebra: A Technological Approach.* Dedham, Mass.: Janson Publications, 1995.

Froehlich, Gary, Kevin G. Bartkovich, and Paul A. Foerster. *Connecting Mathematics. Curriculum and Evaluation Standards for School Mathematics* Addenda Series, Grades 9–12. Reston, Va.: National Council of Teachers of Mathematics, 1991.

Interactive Physics II. San Francisco, Calif.: Knowledge Revolution, 1992. Software.

Jackiw, Nicholas W. The Geometer's Sketchpad. Berkeley, Calif.: Key Curriculum Press, 1991. Software.

Kibbey, David, and Sharon Dugdale. Green Globs and Graphing Equations. Pleasantville, N.Y.: Sunburst Communications, 1988. Software.

Maple. Pacific Grove, Calif.: Brooks/Cole Publishing Co., 1988. Software.

Mathematica. Champaign, Ill.: Wolfram Research, 1991. Software.

Mathematics Exploration Tool Kit. Atlanta: IBM–Eduquest Software, 1989. Software.

Meiring, Steven P., Rheta N. Rubenstein, James E. Schultz, Jan de Lange, and Donald L. Chambers. *A Core Curriculum: Making Mathematics Count for Everyone. Curriculum and Evaluation Standards for School Mathematics* Addenda Series, Grades 9–12. Reston, Va.: National Council of Teachers of Mathematics, 1992.

Microsoft Excel. Redmond, Wash.: Microsoft Corp., 1992. Software.

National Council of Teachers of Mathematics. *Curriculum and Evaluation Standards for School Mathematics.* Reston, Va.: The Council, 1989.

———. *Historical Topics for the Mathematics Classroom.* Reston, Va.: The Council, 1989.

New American Eating Guide. Washington, D.C.: Center for Science in the Public Interest, 1983.

Pollak, Henry. Talk given at the Mathematical Sciences Education Board. Frameworks Conference, Minneapolis, Minn., May 1987.

Sarmiento, Jorge L. "Thousand-Year Record Documents Rise in Atmospheric CO_2." *Chemical and Engineering News.* 31 May 1993: 30–43.

Schwartz, Judah L. The Function Family Register. Pleasantville, N.Y.: Sunburst Communications, 1991. Software.

Silverman, Harold M., Joseph A. Romano, and Gary Elmer. *The Vitamin Book: A No-Nonsense Consumer Guide.* New York: Bantam Books, 1985.

Zbiek, Rose M., and M. Kathleen Heid. "The Skateboard Experiment: Mathematical Modeling for Beginning Algebra." *Computing Teacher* 18 (1990): 32–36.

◆　　◆　　◆　　◆　　◆　　◆　　◆　　◆

APPENDIX: ACTIVITY SOLUTIONS

Activity 1: The Effects of Monomial Terms on Polynomial Functions

1. $f(x) = x^4 - 3x^2$. A function that seems to fit the given curve fairly well, but does not have integer coefficients, has the rule

$$f(x) = x^2(x - 1.75)(x + 1.75).$$

2. $k(c) = -2c^3 + 5c^2$

3. $h(e) = e^5 - 3e^2$

4. $j(b) = b^5 - 4b^3 + b^2$

Activity 2: Oil and Water Don't Mix

The following answer is drawn from pages 95–97 of the Teacher Notes for the *Core-Plus Mathematics Project.*

The relationships are generally linear. Substances with greater density yield graphs with steeper slopes.

1. The slope tells the rate at which mass increases per unit increase in the volume. The density is the mass per unit volume of a substance.

2. Oil is less dense. It rises to the surface of the water, a more dense substance.

3. The value of *a* is the density. The value of *b* gives the mass of the empty container. A model with 0 for *b* could be obtained if the mass of the container were subtracted from the total mass in every measurement.

4. We will use a model of $y = ax + b$, where mass, *y*, is a function of the volume, *x*, of a substance. The values of *a* and *b* will vary, depending on the substance and the mass of the container. If the mass of the container is subtracted from the total mass in each measurement, then $b = 0$ and the model becomes $y = ax$. The equations and inequalities to be solved would be the following:

   (*a*) $y = a(50) + b$

   (*b*) $250 = ax + b$

   (*c*) $ax + b \geq 500$

5. (*a*) Objects that float in water are less dense than water. Objects that sink in water are more dense than water.

   (*b*) The graph of (*volume, mass*) data for lead would have a greater slope than the (*volume, mass*) data for wood.

6. Coke is soluble in water. Even though Diet Coke is less dense than water, it will not form a Coke slick because it will intermingle with the water molecules.

Activity 3: Window Moldings

Note: these answers are approximate.

1. $P(L, W) = 2L + 2W$

♦   ♦   ♦   ♦   ♦   ♦   ♦   ♦

2. (a) Approximately 13.5 feet. If the molding has a constant width of w feet, then the amount of molding needed is $(8w + 13.5)$ feet, provided that the orientation of the molding matters.

(b) Many answers are possible. If the width of the molding is not considered, the window could be 10 feet by 4 feet. If the molding has a constant width of w feet, then (length of window in feet) + (width of window in feet) + $4w = 14$.

(c) If the width of the rectangular window remains constant, increasing the length of the window by x feet would result in an increase of $2x$ feet of molding.

(d) Ignore the width of the molding and assume the window is a regular hexagon. Instead of $P_1(a, b) = 2a + 2b$, the function would be $P_2(s) = 6s$, where s is the length of a side. Assume that the molding is of constant width w for the window. The outside perimeter adds 8 times the width of the molding when the window is rectangular ($P = 2a + 2b + 8w$) and about 7 times the width of the molding when the window is hexagonal:

$$P = 6s + 2\frac{w}{\sqrt{3}}(6) = 6s + 4w\sqrt{3} \approx 6s + 7w.$$

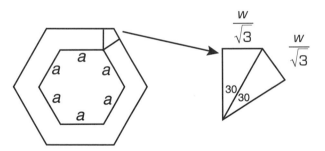

3. The model is a good approximation for the amount of molding. More information is needed for a better model. A possible piece of missing information is the width of the molding. A standard size for molding around a large window may be 1 inch by 4 inches (1 inch thick by 4 inches wide). In reality, this lumber would probably be found to be 3/4 inches by 3 1/2 inches.

Activity 4: Pet Wards

1. The function P describes the number of wall panels as a function of the number of wards, n, with rule $P(n) = 2 + 5(n/2) = (5/2)n + 2$. An equivalent rule can be derived by using a variety of counting methods. For example, counting the east-west walls first and then adding them to the north-south walls results in the rule

$$P(n) = 3\frac{n}{2} + 2\left(\frac{n}{2} + 1\right) = \frac{5}{2}n + 2.$$

Counting the walls on the perimeter, adding the interior north-south walls, and finally adding the interior east-west walls yield

$$P(n) = 2\left(\frac{n}{2}\right) + 4 + 2\left(\frac{n}{2} - 1\right) + \frac{n}{2}.$$

2. Answers will vary.

3. Gather evidence about whether the rules are "the same" or "equiva-lent" by examining graphs and tables of values for the rules. Two function rules will be equivalent if their graphs appear identical for the same viewing window and if their tables give identical output val-ues for the same input values. Examining tables and graphs, howev-er, may be misleading. Graphs and tables may not show function val-ues that are not the same. To know for sure, you will need to apply algebraic rules for producing equivalent forms.

Activity 5: Fuel Bills
See pages 52–55.

Activity 6: Exploring Linearity

1.
Time (in hours)	1	2	3	4	5	6
Charge (in dollars)	3.75	7.50	11.25	15.00	18.75	22.50

2.

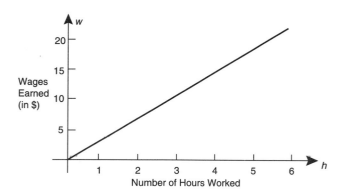

3. $w(h) = 3.75h$

4. $4.75h > 25$. You must work more than 5.26 hours. To the minute, you must work at least 5 hours and 16 minutes.

5. (a) $w_2(h) = 4.35h + 50$

6. To earn $225 in pay, it is necessary to work just over 40 hours (about 40.23 hours, or about 40 hours and 14 minutes).

7. (a) $w(h) = 6.75h + 50$

   (b) You will work fewer hours than your cousin to purchase a CD player. It takes you about 18.2 hours to earn $129 versus your cousin's approximate 20.6 hours of work to earn $189. However, it is unclear how many yard-work jobs are available in each area. Another factor might be differences in weather condi-tions in each area.

   (c) Even increasing your cousin's hourly wage to $7.25 may not help the situation if no work is available. The number of hours needed would decrease by only about 1.42 hours.

(*d*) Using tables or graphs will give approximate values. This approach may be suitable if the workers are paid in increments, such as quarter hours, that are easily displayed in tables or graphs.

8. One possibility: Subtract the two rates of pay ($6.75 – $3.75 = $3.00) then divide $54.00 by this difference. The result, which in this example is 18, is the difference in the number of hours worked.

Activity 7: Exploring the Graphs of Exponential Functions

1. (*a*) $f(x) = 4(2.3)^x$, $f(x) = 2.3^x$, and $f(x) = 8(2.3)^x$

   (*b*) $f(x) = 0.6^x$, $f(x) = 8(0.6)^x$, and $f(x) = 4(0.6)^x$

   (*c*) Since $0 < a < 1$, raising the function to a negative exponent gives a result greater than 1, and the function will increase; raising the function to a positive exponent gives a result less than 1, and the function will decrease.

2. Any nonzero number raised to the zero power is 1; thus both functions are equal when $x = 0$ and $a > 0$.

3. C affects the value of $f(0)$ but not the rate of change.

4. (*a*) $0 < a < 1$, $C > 0$

   (*b*) $a > 1$, $C > 0$

5. If a is between 0 and 1 and $C > 0$, the function decreases. If $a > 1$ and $C > 0$, the function increases.

6. The C-value determines the value of $f(0)$. Also, C seems to cause the graphs to "shift" left or right.

7. Possible answer: All three are alike in that they are exponential functions with $C > 0$. They differ in that in $f(x)$, $0 < a < 1$, whereas in $g(x)$ and $h(x)$, $a > 1$.

8. (*a*) $g(x) = b^x$

   (*b*) $g(x) = b^x$

9. Assuming the a-values are equal in both functions, the graphs will have the same shape, but $h(x)$ will appear to shift to the left of $j(x)$.

Activity 8: Exploring Rational-Function Graphs

1. In each case, if the input is negative, the output is negative, and vice versa. However, as the inputs approach 0 from the negative side, the outputs become very small. Outputs are very large when the input is positive and near 0, but decrease as the inputs continue to increase.

2. The maximum outputs in the table seem to get larger and larger, and the minimum outputs in the table seem to get smaller and smaller; the functions have no minimum or maximum output values.

3. No.

4. As the input values get very close to O, the outputs get either very large or very small depending on whether the input is positive or negative. If O is an input, the functions are undefined.

5. As *a* increases, the functions increase or decrease more rapidly.

6. (*a*) As *x* gets large, the outputs get close to O. If the input is negative, the output becomes very small as *x* approaches O. If the input is positive, the output becomes very large as *x* approaches O.

   (*b*) No. The graphs are not straight lines.

   (*c*) The graphs do not seem to "bow" into the axes as much when the numerators get larger.

Activity 9: Circumference and Area

2. No. The total area of the figure is given by the function

$$A(r) = 2\pi r^2 - 25r + \frac{625}{4\pi},$$

   where *r* = the radius of Circle 1. Since this equation is quadratic, the rate of change is not constant.

Table for questions 3 and 4:

Radius of Circle 1	Area of Circle 1	Sum of Radii	Area of Circle 1 + Area of Circle 2
0.74	1.72	3.99	34.82
0.99	3.07	3.99	31.31
1.23	4.79	3.99	28.25
1.48	6.90	3.99	26.61
1.76	9.77	3.99	25.29
2.01	12.70	3.99	24.96
2.26	16.01	3.99	25.40
2.50	19.71	3.99	26.61
2.75	23.79	3.99	28.58
3.00	28.25	3.99	31.31
3.25	33.09	3.99	34.82

3.

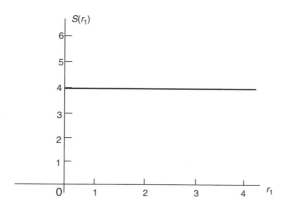

See the foregoing table. The function rule could be $S(r_1) = 3.99$, where *r* is the radius of Circle 1. The rule works because the distance between the centers of the two circles remains constant.

4.

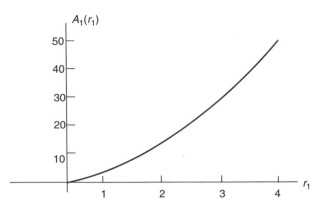

See the foregoing table. If a function fitter is used, the function may resemble

$$A_1(r_1) = 3.098r^2 + 0.1546r - 0.1042.$$

Note the resemblance to $A = \pi r^2$.

5. The distance between the centers of the circles is a constant, so the sum of the radii is constant. Hence, the first radius is a linear function of the second. The graph of a constant function is linear. The measurement of area requires squaring a linear function, resulting in a quadratic function.

Activity 10: Similarities and Differences in Properties of Different Families of Functions

Function Form	Rate of Change	Symmetry Feature	No. of Max/ Min Values	Special Features
$mx + b$	Constant	None	None	Graphs are lines.
a^x	Variable	None	None	Graphs are not straigtht and are either increasing or decreasing but not both. Always passes through (0, 1).
$ax^2 + bx + c$	Variable	About a vertical line through the vertex	1	Graphs are parabolas and have one piece that "turns" (switching between increasing and decreasing) once.
$ax^3 + bx^2 + cx + d$	Variable	None	0, or 2 Relative Extrema	Graphs are one piece. Range of output values is all real numbers.
$ax^4 + bx^3 + cx^2 + dx + e$	Variable	None	1, or 3 Relative Extrema	Graphs are one piece.
$\dfrac{a}{x}$	Variable	About the origin	None	Asymptotes. Graphs have two "pieces," one above and one below the input-axis.

$\dfrac{a}{x^2}$	Variable	About the output-axis	None	Asymptotes
				Graphs have two "pieces," one left of and one right of the output-axis and both on the same side of the input-axis.
$\dfrac{a}{x^3}$	Variable	About the origin	None	Asymptotes
				Graphs have two "pieces," one above and one below the x-axis.
$\dfrac{a}{x^4}$	Variable	About the output-axis	None	Asymptotes
				Graphs have two "pieces," one left of and one right of the output-axis and both on the same side of the input-axis.

Activity 11: Applications of Polynomial Functions

1. Possible answer: $9.00, because 9(119) produces the most money. See the following table.

Ticket Price (in $)	Average Number of Tickets Sold	Income from Admissions (in $)
5	158	790
7	142	994
9	119	1071
11	97	1067

   Additional information that might be needed: Does the cost change with greater sales? Could other activities in the community affect participation? Is the goal greater attendance or more money (revenue or profit)?

2. A function fitter will find many different close fits to the data. A possible argument for a linear function is that as the price increases, the attendance will decrease. A graph seems to indicate that this decrease looks like a straight line. However, some possible function rules follow:

   Linear: $N(x) = -10.3x + 211.4$; Goodness-of-fit (GOF): 99.714

   Quadratic: $N(x) = -0.375x^2 - 4.300x + 189.276$; GOF: 99.925

   Exponential: $N(x) = 244.499(0.921^x)$; GOF: 99.097

   Linearity seems reasonable because demand is often well modeled by a linear function, because the data fit the curve very well, and because the goodness-of-fit measure is very high.

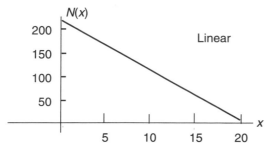

3.

Price ($)	Profit ($)	Price ($)	Profit ($)	Price ($)	Profit ($)
0	−800	9	352	17	288
1	−608	10	400	18	208
2	−432	11	432	19	112
3	−272	12	448	20	0
4	−128	13	448	21	−128
5	0	14	432	22	−272
6	112	15	400	23	−432
7	208	16	352	24	−608
8	288			25	−800

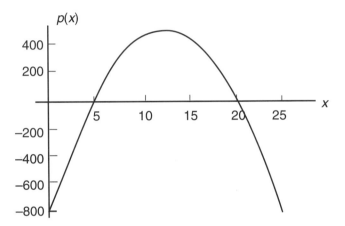

(a) As the prices increase from $0 to $25, the profit increases and then decreases. In fact, the profit is negative at extreme prices in this interval. The profits shown in the table increase up to $448 then decrease.

(b) Table: Break-even points occur at ticket prices of $5 and $20. The table seems better at telling particular values.

Graph: The relation is parabolic in nature. The graph quickly tells us the overall pattern—increase rapidly and then more slowly and then decrease slowly and then more rapidly.

4. The break-even ticket prices are $5 and $20. The ticket price to maximize profit is between $12 and $13.

5. To reflect revenue earned from concession sales would require taking the profit function rule $p(x)$, adding to it the revenue function $r(x) = xN(x)$ from the concession sales, and subtracting from it the cost function for the concession sales. However, we assume that the cost of food is already included in the $435.00 daily operating costs. In this situation, we use $3.50 times the number of tickets as the revenue from the concessions, which yields the revised profit function, pr:

$$pr(x) = (3.5 + x)(-8x + 200) - 800 = -8x^2 + 172x - 100.$$

For the following analysis, we will assume a demand function of $N(x) = -8x + 200$.

Price ($)	Estimated Number of Tickets Sold	Revenue from Concession Sales ($)	Profit (including concession revenue) ($)	Profit (without concession revenue) ($)
0	200	700	−100	−800
1	192	672	64	−608
2	184	644	212	−432
3	176	616	344	−272
4	168	588	460	−128
5	160	560	560	0
6	152	532	644	12
7	144	504	712	208
8	136	476	764	288
9	128	448	800	352
10	120	420	820	400
11	112	392	824	432
12	104	364	812	448
13	96	336	784	448
14	88	308	740	432
15	80	280	680	400
16	72	252	604	352
17	64	224	512	288
18	56	196	404	208
19	48	168	280	112
20	40	140	140	0
21	32	112	−16	−128
22	24	84	−188	−272
23	16	56	−376	−432
24	8	28	−580	−608
25	0	0	−800	−800

$$pr : x \rightarrow -8 \cdot x^2 + 172 \cdot x - 100$$

$$p : x \rightarrow -8 \cdot x^2 + 200 \cdot x - 800$$

$$pr : x \rightarrow -8 \cdot x^2 + 172 \cdot x - 100$$

$$rc : x \rightarrow -28.0 \cdot x + 700.0$$

$$p : x \rightarrow -8 \cdot x^2 + 200 \cdot x - 800$$

Both the table and the graph verify the new profit function, since they show that $pr(x) = rc(x) + p(x)$. In other words, the values of the new profit function can be obtained by adding the values of the original profit function to the corresponding values of the concession-revenue function. (Note that the concession-revenue function rule was derived from the demand function, $N(x) = -8x + 200$; that is, $rc(x) = 3.5(-8x + 200)$.

Including the concession revenue seems to change the price that yields maximum profit. This situation is possible because the concession revenue from the extra attendance may have offset the smaller profit that results from lower admission prices.

6. A ticket price of $12.50 would maximize profit when concession revenue is not included. A ticket price of about $10.75 would maximize profit when concession revenue is included. A ticket price of $0 would maximize the number of people attending at 200.

Activity 12: Using Functions of Many Variables to Analyze Dietary Choices

1. The given menu, although lower in calories than the desired amount, exceeds the required amount of protein, vitamin C, calcium, and iron. The amount of sodium consumed is at the high end of the desired range.

2–4. A possible menu is given below:

Item	Calories	Protein (g)	Vit. C (mg)	Iron (mg)	Sodium (mg)
Breakfast:					
1/2 grapefruit	41	0.5	38	0.4	1
3 pancakes	183	5.7	T	0.9	456
1/4 c. maple syrup	198	0	0	0.9	8
1 c. skim milk	88	8.8	2	0.1	127
Lunch:					
2 sandwiches:					
4 sl. wheat bread	244	10.4	T	3.2	528
2 Tbs. jam	108	0.2	T	0.4	4
2 Tbs. peanut butter	188	8	0	0.6	194
1 c. chicken noodle soup	62	3.4	T	0.5	950
1 c. applesauce	232	0.5	2	1.2	5
2 c. skim milk	176	17.6	4	0.2	254
Dinner:					
6 oz. broiled swordfish	276	44	0	1.8	0
1 Tbs. tartar sauce	74	0.2	T	0.1	99
1/4 lb. french fries	311	5	24	1.5	7
1 c. spinach	48	5.8	56	5	98
1 med. lettuce salad	28	2.4	16	4	18
2 Tbs. 1000 island dressing	160	0.2	T	0.2	224
1 c. skim milk	88	8.8	2	0.1	127
Total	2505	121.3	144	21.1	3100

◆   ◆   ◆   ◆   ◆   ◆   ◆   ◆

Activity 13: Refreshment-Stand Profits

1. The graph of the survey data along with a best-fitting linear-function rule appears below.

$$M(t) = -319.6t + 4002.2$$

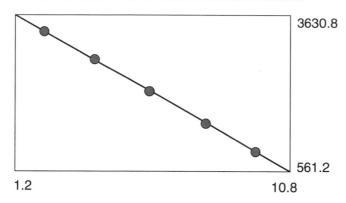

F1 : x → $-\dfrac{1598}{5} \cdot x + \dfrac{20011}{5}$

3630.8

561.2

1.2   10.8

InData	OutData
2	3375
4	2706
6	2095
8	1430
10	817

2. $P_1(a) = at - 12(110)$, where t is the price of a ticket in dollars.

P : t → t • $\left[-\dfrac{1598}{5} \cdot t + \dfrac{20011}{5} \right] - 1320$

Price (in $)	Number of People	Profit (in $)
3.00	3043.4	7 810.20
4.50	2564.0	10 218.00
6.00	2084.6	11 187.60
7.50	1605.2	10 719.00
9.00	1125.8	8 812.20
10.50	646.4	5 467.20
12.00	167.0	684.00

3. The profit function is shown in the answer to question 2. It is equivalent to $P(t) = -319.6t^2 + 4002.2t - 1320$.

4. From the table in the answer to question 2, it seems that the maximum profit would occur if the price is about $6.00. We could improve the estimate with another table. From this table, it seems that the maximum profit would occur if the price is about $6.30. We could do more tables, but ticket prices are rarely even multiples of nickels or dimes!

Price (in $)	Number of People	Profit (in $)
5.50	2244.40	11 024.200
5.60	2212.44	11 069.664
5.70	2180.48	11 108.736
5.80	2148.52	11 141.416
5.90	2116.56	11 167.704
6.00	2084.60	11 187.600
6.10	2052.64	11 201.104
6.20	2020.68	11 208.216
6.30	1988.72	11 208.936
6.40	1956.76	11 203.264
6.50	1924.80	11 191.200
6.60	1892.84	11 172.744

Activity 14: Exploring Function Orbits

1. If $|x_0| > 1$, the orbit grows without bound. If $|x_0| < 1$, the orbit approaches 0. If $|x_0| = 1$, the orbit is constant.

2. When the input value is in radians, the orbit approaches 0.739 085 133 (approximately). When the input value is in degrees, the orbit approaches 0.998 477 415 (approximately).

3. For $h(x)$ where $x_0 > 0$, the orbit approaches 2. For $h(x)$ where $x_0 < 0$, the orbit approaches –2.

    For $k(x)$ where $x_0 > 0$, the orbit approaches 3. For $k(x)$ where $x_0 < 0$, the orbit approaches -3.

    For $m(x)$ where $x_0 > 0$, the orbit approaches $\sqrt{2}$. For $m(x)$ where $x_0 < 0$, the orbit approaches $-\sqrt{2}$.

    For

    $$\frac{x + \dfrac{a^2}{x}}{2},$$

    the orbits are constant a if $x_0 = a$, $x_0 \neq 0$, and $a \leq 0$. Otherwise, if $x_0 > 0$, orbits approach a; if $x_0 < 0$, orbits approach $-a$.

2.

$$X = \begin{bmatrix} 0 & 1 & 1 & 0 \\ 0 & 0 & 1 & 1 \end{bmatrix}$$

X represents the vertices of the unit square.

$$AX = \begin{bmatrix} 0 & 1 & 1 & 0 \\ 0 & 0 & -1 & -1 \end{bmatrix}$$

A reflects the unit square over the horizontal axis.

$$BX = \begin{bmatrix} 0 & -1 & -1 & 0 \\ 0 & 0 & -1 & -1 \end{bmatrix}$$

B rotates the unit square 180 degrees about the origin.

$$CX = \begin{bmatrix} 0 & -1 & -1 & 0 \\ 0 & 0 & 1 & 1 \end{bmatrix}$$

C reflects the unit square over the vertical axis.

$$DX = \begin{bmatrix} 0 & 0 & 1 & 1 \\ 0 & 1 & 1 & 0 \end{bmatrix}$$

D reflects the unit square over the line $y = x$.

$$EX = \begin{bmatrix} 0 & 1 & 0 & -1 \\ 0 & -1 & -1 & 0 \end{bmatrix}$$

E reflects the unit square over the line $y = -x$ then maps the result using the shear transformation

$$\begin{bmatrix} -1 & 1 \\ 1 & 0 \end{bmatrix}.$$

$$FX = \begin{bmatrix} 0 & 1 & 0 & -1 \\ 0 & 1 & 2 & 1 \end{bmatrix}$$

F rotates the unit square 45 degrees counterclockwise and dilates it by $\sqrt{2}$.

$$GX = \begin{bmatrix} 0 & 0 & 1 & 1 \\ 0 & -1 & -1 & 0 \end{bmatrix}$$

G rotates the unit square 90 degrees clockwise.

$$HX = \begin{bmatrix} 0 & 0 & -1 & -1 \\ 0 & 1 & 1 & 0 \end{bmatrix}$$

H rotates the unit square 90 degrees counterclockwise.

3. Isometries: *A*, *B*, *C*, *D*, *G*, and *H*

4. The fixed points are—

   A   Points on the horizontal axis

   B   (0, 0)

   C   Points on the vertical axis

   D   Points on the line $y = x$

   E   (0, 0)

   F   (0, 0)

   G   (0, 0)

   H   (0, 0)

5.

$$ABX = \begin{bmatrix} 0 & -1 & -1 & 0 \\ 0 & 0 & 1 & 1 \end{bmatrix}$$

AB reflects the unit square over the vertical axis.

$$BCX = \begin{bmatrix} 0 & 1 & 1 & 0 \\ 0 & 0 & -1 & -1 \end{bmatrix}$$

BC reflects the unit square over the horizontal axis.

$$DFX = \begin{bmatrix} 0 & -1 & -1 & 0 \\ 0 & 0 & 1 & 1 \end{bmatrix}$$

DF reflects the unit square over the horizontal axis, rotates it 45 degrees counterclockwise, and dilates it by $\sqrt{2}$.

$$EFX = \begin{bmatrix} 0 & 0 & -2 & -2 \\ 0 & -1 & 0 & 1 \end{bmatrix}$$

EF reflects the unit square over the line $y = -x$, stretches the resulting figure so that *y*-values stay fixed and *x*-values move to twice their value, and finally maps the result using the shear transformation

$$\begin{bmatrix} 1 & 0 \\ -0.5 & 1 \end{bmatrix}.$$

$$FEX = \begin{bmatrix} 0 & 2 & 1 & -1 \\ 0 & 0 & -1 & -1 \end{bmatrix}$$

FE reflects the unit square over the *x*-axis, stretches the resulting figure so that *y*-values stay fixed and *x*-values move to twice their value, and finally maps the result using the shear transformation

$$\begin{bmatrix} 1 & 1 \\ 0 & 1 \end{bmatrix}.$$

Activity 16: What Does F(S) = KS Do to the Area of the Unit Square S?

1. (a) The area of the enclosing rectangle is (3 – (–2))(3 – 0) = 15.

   (b) The area of A is (1/2)(1)(3) = 1.5.
      The area of B is (1/2)(2)(2) = 2.
      The area of C is (1/2)(2)(2) = 2.
      The area of D is (1/2)(1)(3) = 1.5.

   (c) The area of quadrilateral S' is 15 – (1.5 + 2 + 2 + 1.5) = 8.

   (d) The area of S' is the absolute value of the determinant of K.

2. If R is a rectangle, then the area of R' is the value of the determinant of K times the area of R. For example, if the area of R is 3, then the area of R' is three times the absolute value of the determinant of K.

3. If P is a parallelogram, the result will be the same as that in question 2.

4. The area of KQ will be the area of Q times the absolute value of the determinant of K.

Activity 17: Equivalent Forms I

Let n represent the number of matchsticks on one side of the figure.

1. Procedure 1: Count the north-south matchsticks and add the result to the number of east-west matchsticks.

   Counting the north-south matchsticks first, we get n(n + 1). Counting the east-west matchsticks, we get n(n + 1). The total number of matchsticks is n(n + 1) + n(n + 1).

   Procedure 2: Look at each design as a column of unfinished squares and a column of matchsticks to finish the right side:

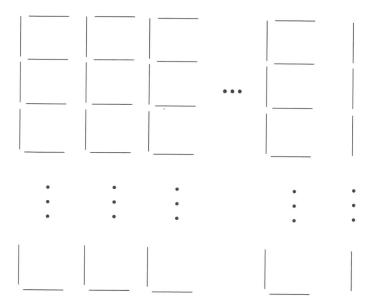

Each column of unfinished squares consists of n vertical and $n + 1$ horizontal matchsticks. The total number of matchsticks is $n(n + (n + 1)) + n$.

2. Graphs and tables of the two functions seem to give the same output values for given input values.

Activity 18: Equivalent Forms II

1. (a)

$$\frac{1}{2}\left(BF + CG\right)AC + AB^2 - \frac{1}{2}\left(\pi\left(\frac{CB}{2}\right)^2\right)$$

But

$$BF + CG = (BA + AF) + AC = AB + 2AC$$

and

$$CB = \sqrt{AC^2 + AB^2}.$$

So the original expression, as a function of AB and AC, is

$$\frac{1}{2}\left(AB + 2AC\right)AC + AB^2 - \frac{1}{2}\left(\pi\left(\frac{\sqrt{AC^2 + AB^2}}{2}\right)^2\right).$$

(b)

$$AC^2 + AB^2 + \frac{1}{2}AB \cdot AC - \frac{1}{2}\left(\pi\left(\frac{\sqrt{AC^2 + AB^2}}{2}\right)^2\right)$$

(c)

$$\left(AC^2 - \frac{1}{2}\pi\left(\frac{AC}{2}\right)^2\right) + \left(AB^2 - \frac{1}{2}\pi\left(\frac{AB}{2}\right)^2\right)$$

(d)

$$\left(AC^2 + AB^2\right) - \left(\frac{1}{2}\pi\left(\frac{\sqrt{AC^2 + AB^2}}{2}\right)^2 - \frac{1}{2}AB \cdot AC\right)$$

The expressions in (a), (b), and (d) are equivalent and equal to the required area.

2. (a) $\frac{1}{2}\left(\frac{1}{2}AC\right)\left(\frac{1}{4}AK\right) + \frac{1}{2}\left(\frac{1}{2}AK\right)\left(\frac{1}{4}AC\right)$

$$+ \frac{1}{2}\left(\frac{1}{2}AC\right)\left(\frac{1}{4}AK\right) + \frac{1}{2}\left(\frac{1}{2}AK\right)\left(\frac{1}{4}AC\right)$$

(b)

$$\left(\frac{1}{2}(AC)(AK)\right) - \left(\left((AC)(AK)\right) - \left(\left(\frac{1}{2}AK\right)\left(\frac{1}{2}AC\right)\right)\right)$$

(c)

$$2\left(\left(\frac{1}{2}AK\right)\left(\frac{1}{2}AC\right)\right) - \left(\left(\frac{1}{2}AK\right)\left(\frac{1}{2}AC\right)\right)$$

(d)

$$\left(AC\right)\left(AK\right) - \frac{1}{2}\left(AC\right)\left(AK\right) + \left(\left(\frac{1}{2}\left(AC\right)\left(AK\right)\right) + \left(\left(\frac{1}{2}AK\right)\left(\frac{1}{2}AC\right)\right)\right)$$

The expressions in (a) and (c) are equivalent and equal to the required area.

3. Given segment $AD \perp$ segment AC.

Method 1: We will use the fact that $\triangle JFE$ and $\triangle JHG$ are congruent, with $JH = (1/2)JC$.

Then the desired area is twice the area of $\triangle JHG$, or

$$2\left(\frac{1}{2}\right)^2\left(\frac{1}{2}(JC)(CB)\right) = 2\left(\frac{1}{2}\right)^2\left(\frac{1}{2}(AC)(AD)\right).$$

Method 2: From the given information, we have $EF = (1/2)AD$ and $HG = (1/2)BC = (1/2)AD$. Also, $\overline{AC} \perp \overline{AD}$ implies $\overline{JF} \perp \overline{EF}$ and $\overline{JH} \perp \overline{HG}$, since $\overline{EF} \parallel \overline{AD} \parallel \overline{CB} \parallel \overline{HG}$. The desired area is

$$\text{Area}\left(\triangle JFE\right) + \text{Area}\left(\triangle JHG\right) = \frac{1}{2}\left(EF\right)\left(JF\right) + \frac{1}{2}\left(HG\right)\left(JH\right)$$

$$= \frac{1}{2}\left(\frac{1}{2}(AD)\right)\left(\frac{1}{2}(JA)\right) + \frac{1}{2}\left(\frac{1}{2}(CB)\right)\left(\frac{1}{2}(JC)\right)$$

$$= \frac{1}{2}\left(\frac{1}{2}(AD)\right)\left(\frac{1}{2}\left(\frac{1}{2}(AC)\right)\right)$$

$$+ \frac{1}{2}\left(\frac{1}{2}(AD)\right)\left(\frac{1}{2}\left(\frac{1}{2}(AC)\right)\right).$$

ANNOTATED LIST OF TECHNOLOGY RESOURCES

Catalysts for Concept Development

Vectors

de Lange, Jan. Ballooning. Pleasantville, N.Y.: Sunburst/Wings for Learning, 1993.
Macintosh (1MB)

Through the simulation of flying a hot-air balloon, students can develop their concepts of direction, magnitude, angles, and vector addition. The student has the opportunity to plan different routes and notice the related wind effects.

Functions, graphs, and properties of functions

Dugdale, Sharon, and David Kibbey. Green Globs and Graphing Equations. Pleasantville, N.Y.: Sunburst/Wings for Learning, 1986.
Apple, IBM, Tandy 1000 (128 K)

This program offers environments, including games, in which students manipulate and explore the graphs of functions. For example, Green Globs is a game in which students shoot "green globs" by specifying function rules for function graphs that pass through as many targets as possible. Tracker is a game in which students use horizontal and vertical probe shots to determine the equations of hidden graphs.

————. Interpreting Graphs. Pleasantville, N.Y.: Sunburst/Wings for Learning, 1985.
Apple, IBM, Tandy 1000 (256 K)

In one program, students match graphs to physical events. In another program, the game of Escape, students interpret various graphical representations in their attempts to set blockades for escaping robbers.

Dugdale, Sharon, David Kibbey, and Linda Wagner. Slalom, ZOT, and ZigZag: Challenges in Graphing Equations. Pleasantville, N.Y.: Sunburst/Wings for Learning, 1991.
IBM, Tandy 1000 (256 K)

Through a series of games, students work with polynomial, rational, and absolute-value functions. They construct functions with given zeros, work with rational functions creating desired asymptotes, explore absolute-value functions, and work with equations whose graphs are pairs of parallel or perpendicular lines.

Harvey, Wayne, Judah Schwartz, and Michal Yerushalmy. The Function Supposer: Explorations in Algebra. Pleasantville, N.Y.: Sunburst/Wings for Learning, 1990.
IBM, Tandy 1000 (512 K)

This program allows students to use graphs to explore relationships among functions, operations on functions, and construction of functions. The Function Supposer allows students to investigate the graphical effects of binary operations on functions. For example, after giving function rules for f and g and designating $h(x) = f(x) \cdot g(x)$, students can investigate the effects on $h(x)$ of various transformations (e.g., translating, stretching, flipping) of f and g.

————. Visualizing Algebra: The Function Analyzer. Pleasantville, N.Y.: Sunburst/Wings for Learning, 1992.
IBM, Tandy 1000 (512 K)

Through linked representations, students explore relationships among symbolic, graphical, and numerical representations. Students see how modifying a function by changing its graph or symbolic representation affects other representations of the function.

Schwartz, Judah. The Function Family Register. Pleasantville, N.Y.: Sunburst/Wings for Learning, 1991. (Not currently being distributed by Sunburst/Wings for Learning.)
IBM, Macintosh

With the Function Family Register, students can explore two-parameter families of functions by manipulating the parameters represented as ordered pairs and watching the effect on the function graphs. They can choose from among different types of functions (polynomial, trigonometric, transcendental, or rational functions) or create their own function family. The program can also generate a target function for students to match.

————. un*Solving. Pleasantville, N.Y.: Sunburst/Wings for Learning, 1992.
IBM, Tandy 1000, Macintosh

The software un*Solving provides opportunities for students to enhance their understanding of the meaning of "solving" an equation or inequality. The program can give the student a solution set for which the user must find a matching equation or inequality, and it can give a linear or quadratic equation or inequality to solve. In the latter situation, the student specifies transformations that produce equivalent equations or inequalities.

Schwartz, Judah, and Michal Yerushalmy. The Function Supposer: Symbols and Graphs. Pleasantville, N.Y.: Sunburst/Wings for Learning, 1992.
Macintosh (2.5 MB), IBM, Tandy 1000 (512 K)

This program allows students (1) to plot functions, generate tables of values, and find roots; (2) to see symbolic and graphical representations for factoring, expanding, simplifying, and collecting terms; and (3) to explore the symbol-graph connection for equations, inequalities, and identities.

Tools of Mathematics: Advanced Algebra from MacNumerics II. Acton, Mass.: William K. Bradford Publishing Co., 1993.
Macintosh

This program furnishes tools to help the study, understanding, learning, and teaching of algebra topics through visual and intuitive exploration. The tools in this package include ones designed for studying conic sections, inequalities, linear systems, and polynomials.

Yerushalmy, Michal, and Beba Shternberg. Algebraic Patterns from Arithmetic to Algebra. Pleasantville, N.Y.: Sunburst/Wings for Learning, 1994.
Macintosh, IBM, Tandy 1000

Algebraic Patterns allows students to explore lattices composed of arithmetic series to learn to detect and generalize numerical patterns. The program addresses such concepts as equation writing, relationships between quantities, and properties of integers.

————. The Algebra Sketchbook. Pleasantville, N.Y.: Sunburst/Wings for Learning, 1994.
IBM, Tandy 1000

In the Algebra Sketchbook, students link the characteristics of a function with an appropriate graph and the characteristics of a graph with a matching situation.

Tool for Mathematical Modeling

Edwards, Lois. Data Models. Pleasantville, N.Y.: Sunburst/Wings for Learning, 1991.
Macintosh

Data Models is a mathematical-modeling tool. Students can enter data pairs of their choosing, plot the data or transformed data, and fit a curve (median-fit or least squares) to the plotted points.

Computer-Algebra and Symbolic-Manipulation Tools

Child, J. Douglas. Calculus T/L II. Pacific Grove, Calif.: Brooks/Cole Publishing Co., 1993.
Macintosh

Calculus T/L II is a front-end program operating on the symbolic-manipulation program Maple. It gives the student an interface that facilitates the use of a subset of Maple for generating tables of values, graphing functions of one and two variables, solving equations, producing equivalent expressions, and curve fitting. Students and teachers can use programs in the form of scripts to access symbolic manipulation of different-sized chunks.

Derive. Honolulu, Hawaii: Soft Warehouse, 1988.
IBM

Derive is a computer-algebra program that does exact and approximate arithmetic to thousands of digits; solves equations; performs such symbolic manipulation as simplifying, expanding, and factoring symbolic-algebraic expressions; manipulates vectors and matrices; and does two-dimensional and three-dimensional plotting.

Maple V. Pacific Grove, Calif.: Brooks/Cole Publishing Co., 1980–91.
Macintosh

Maple V is a computer-algebra system whose symbolic-algebra capabilities include finding exact and numerical solutions for a variety of equation types, performing a wide range of symbolic manipulations, defining functions and procedures, plotting a variety of graphs of functions of one or two variables, and operating on matrices.

MathCAD. Cambridge, Mass.: Mathsoft, 1989.
IBM, Macintosh

The MathCAD package is a tool for mathematical calculation that integrates equations, text, and graphics into a single document. Students can define variables and enter formulas anywhere on the screen and can intersperse explanatory text throughout the document. MathCAD formats equations automatically in standard notation and calculates the results instantly.

Mathematica. Champaign, Ill.: Wolfram Research, 1990.
Macintosh, IBM

Mathematica is a computer-algebra system. Its numerical capabilities that might be useful in algebraic settings include the ability to calculate with numbers of any given precision and the ability to display results in rational, decimal, scientific, radical, and complex forms. Its algebraic-symbolic-computation abilities include expanding, factoring, and simplifying polynomials and rational expressions as well as solving classes of solvable polynomial equations and systems of equations. Its graphical capabilities include production of two- and three-dimensional graphs, with user-controlled shading, color, and other attributes. Its high-level programming language supports procedural, functional, and rule-based programming.

Theorist. San Francisco, Calif.: Prescience Corp., 1990.
Macintosh

Theorist is a WYSIWYG (what you see is what you get) symbolic-algebra and graphing program. Its features include point-and-drag access to manipulating symbols and notebooks that offer organized work space for algebraic, numeric, and graphical study.

Simulation Tools

The Geometer's Sketchpad. Berkeley, Calif.: Key Curriculum Press, 1992.
IBM, Macintosh

The Geometer's Sketchpad is a geometric-construction tool that allows the student to construct and manipulate (in a point-and-click environment) two-dimensional geometric figures. The student can request measures for parts of the figure and can ask for calculations involving those measurements (e.g., ratio). He or she can then dynamically change the figure and watch the effects on the related lengths, angle measures, and ratios.

Interactive Physics II. San Francisco, Calif.: Knowledge Revolution, 1989.
Macintosh

Interactive Physics II is a computer-based motion laboratory. Students can choose or design their own experiments, then collect and analyze data related to their experiments. Available simulation elements for experimentation include ropes, motors, pulleys, and springs. Students can vary the object mass, elasticity, friction, and charge.

Integrated Package

f(g) Scholar. Southhampton, Penn.: Future Graph, 1994.
IBM, Macintosh

The software f(g) Scholar is a fully integrated mathematics system that includes a calculator and programming language, an enhanced spreadsheet, a graphing package, and a drawing package.